HIGHER EDUCATION IN THE TWENTY-FIRST CENTURY II

PAPERS FROM THE TWO-DAY WORKSHOP HELD AT AHLIA UNIVERSITY, KINGDOM OF BAHRAIN, 18 & 19 MAY 2014

Higher Education in the Twenty-First Century II

Editors

Abdulla Y.A. Al-Hawaj
Ahlia University, Kingdom of Bahrain

E.H. Twizell
Brunel University London, UK

CRC Press
Taylor & Francis Group
Boca Raton London New York Leiden

CRC Press is an imprint of the
Taylor & Francis Group, an **informa** business

A BALKEMA BOOK

CRC Press/Balkema is an imprint of the Taylor & Francis Group, an informa business

© 2016 Taylor & Francis Group, London, UK

Typeset by MPS Limited, Chennai, India

All rights reserved. No part of this publication or the information contained herein may be reproduced, stored in a retrieval system, or transmitted in any form or by any means, electronic, mechanical, by photocopying, recording or otherwise, without written prior permission from the publishers.

Although all care is taken to ensure integrity and the quality of this publication and the information herein, no responsibility is assumed by the publishers nor the author for any damage to the property or persons as a result of operation or use of this publication and/or the information contained herein.

Published by: CRC Press/Balkema
P.O. Box 11320, 2301 EH Leiden, The Netherlands
e-mail: Pub.NL@taylorandfrancis.com
www.crcpress.com – www.taylorandfrancis.com

ISBN: 978-1-138-02925-5 (Hardback)
ISBN: 978-1-315-64290-1 (eBook PDF)

Table of contents

Introduction	VII
Usefulness of the cash flow statement for Bahraini banks' lending decisions H.A.M. Husain	1
Quotas and empowerment: Enhancing female leadership in corporate boardrooms L.F. Alhalawachi & S. Costandi	15
Evaluating professional development in healthcare with outcome based models L.M. Shibu, E.K. Rajab & T. Eldabi	31
Impact of quality assurance on accountability in policy networks M. Al Oraibi, S. Costandi & T. Eldabi	43
Crisis management and business continuity in the Kuwaiti oil sector M. Al-Tahous	55
Challenges faced by female entrepreneurs in developing countries N.G.R. Taqi	67
Knowledge sharing culture in higher education: Critical literature review O.F.A. Al Kurdi, A. Ghoneim & A. Al Roubaie	75
The Bahraini corporate governance code: Its effect on the corporate sector S.H. Al Hasan	89
Review of energy management policies in healthcare buildings S. Shehab	99
A data mining approach for investigating students' completion rates S. Bhaskaran, K. Lu & M. Al Aali	105
Organizational effectiveness in secondary schools: An empirical study T.H.A. Maki, S. Singh, T. Eldabi & W. Elali	117
Appendix: Students and Supervisors	129
Author index	131

Introduction

Higher Education institutions in the countries of the Gulf Co-operation Council, now, more than at any other time in the past, bear a heavy burden and so are charged with playing a decisive role in the development of the six nations.

Since Ahlia University's inception in 2001, we have taken the lead in promoting excellence and social responsibility in all areas of higher education. Having been the first licensed private university in the Kingdom of Bahrain, Ahlia University has become the hub for private education in the *GCC* region.

Ahlia University is committed to meeting the challenges facing the Kingdom of Bahrain, the wider *GCC* and the world at large. Consequently, the University's research strategy includes providing facilities that encourage and support research students. Ahlia University will contribute to the advancement and propagation of knowledge by encouraging its faculty and research students to publish their original research that is of international significance and can be applied to real-life situations for the benefit of the Kingdom of Bahrain, the wider *GCC* and elsewhere in the world.

Ahlia University's leadership in the domestic, private higher-educational environment is manifest in having the only accredited Ph.D. programme in the Kingdom of Bahrain, in collaboration with Brunel University London in the United Kingdom. Successful students are awarded the Ph.D. degree of Brunel University London.

The first student on the programme registered in 2005 and was awarded her degree in July 2009 at a ceremony held at Brunel University London.

Brunel Business School and Ahlia University offer an exciting, productive and supportive environment for our research students, who experience a thorough exposure to professional research guidance, including techniques, approaches and methods. The central aim is to enable the students to undertake original independent research, leading to the writing of a thesis which is defended at a *viva voce* at Brunel University London.

Students have two supervisors: an Ahlia-based supervisor who is supported by a Brunel-based supervisor. All faculty, at Ahlia University and Brunel University London, who are engaged in this Ph.D. programme, are experienced and respected international researchers who are committed to ensuring our students' success.

Brunel and Ahlia Universities have a strong research culture and faculty are engaged in a variety of projects funded by external bodies such as the UK's Economic & Social Research Council (ESRC) and the European Union. Individual members of staff also have research and consultancy links with a number of leading organizations worldwide. Academic staff members at both universities have published widely in refereed international journals and many have authored or co-authored a number of books.

An important annual feature of the programme is the two-day workshop held at Ahlia University at which all students are expected to give a presentation on their research to date. This is normally attended by all supervisors and other interested members of the faculty of Ahlia University. After the workshop, each participating student is expected to record the contents of his/her presentation in a research paper. This volume contains the papers written by those students who made presentations at the workshop held in 2014.

I would like to take this opportunity to thank all those who have contributed to the success so far of this Ph.D. programme. Professor Zahir Irani, Dean of College of Business, Arts and Social Sciences was, and remains, instrumental in making the programme so successful. He has been helped enormously by Dr Tillal Eldabi, the current Director of the programme. Both have received the support of Professor Chris Jenks, former Vice-Chancellor and Principal of Brunel University London and by Professor Julia Buckingham, the current Vice-Chancellor and President of Brunel.

At Ahlia University, Professor Shawqi Al Dallal, Dean of the College of Graduate Studies & Research, and Dr Samia Costandi, Associate Dean of the College of Graduate Studies & Research, now manage the academic aspects of the programme at Ahlia. These faculty members have received great support from the programme administrators: Mr Gowrishankar Srinivasan and Ms Hessa Al Dhaen at Ahlia University, and Ms Ela Heaney, the CBASS PGR Admin Group, and the College Education Manager Dr Stephen Mullins at Brunel. The students' supervisors at Brunel University London and at Ahlia University are listed in the Appendix. We are extremely grateful to all of these colleagues, who have given the students the benefit of their wisdom and substantial experience.

The ultimate success of the programme depends also on our students. After graduation they will be Ahlia University's diplomats and will give their countries the benefit of the knowledge they have gained and the lessons they have learned, thus enriching the *GCC* countries and helping them to meet the challenges they face as they build the capacity needed to compete within the *GCC* and globally.

The quest for excellence in the continuing development of the higher education sector in the Kingdom of Bahrain goes on and this valuable Ph.D. programme reflects Ahlia University's commitment to promoting research in the Kingdom; this is a major feature of Ahlia's strategic plan.

Abdulla Y.A. Al-Hawaj
Ahlia University, Kingdom of Bahrain

Usefulness of the cash flow statement for Bahraini banks' lending decisions

Hassan A.M. Husain
Ahlia University, Kingdom of Bahrain

ABSTRACT: Accounting and commercial lending studies indicate the importance of the cash flow statement and its information and techniques to account users for better decision making. Till the date of writing no attempts have been made to provide data on the actual use of cash flow information or techniques in lending decisions. This study aims to investigate empirically the impact of understanding, using, and analysing the borrower's cash flow statement on the commercial lending decision and exploring the "cashflowability" information and techniques used by lenders to assess the creditworthiness of the borrowers in order to improve the lending analysis and smooth the lending decision. Thus, we expect that when presented with empirical tasks conducted in the Kingdom of Bahrain and designed to measure the influence of cashflowability on lenders' assessments of credit-worthiness, the lenders would use a cash flow statement and its information and techniques as a relevant signal.

1 INTRODUCTION

The last twenty years (1995–2014) have seen global dramatic losses in the banking industry. Commercial banks in the 1990s and 2000s faced many problems such as increasing competition, increased regulation including capital requirement, higher costs and more sophisticated customers, as well as more problematic borrowers due to the credit exposure that turned bad. The difficulty of collecting the money from some borrowers on its due date is an old problem associated with lending money. The immediate cause of this, in almost all cases, is that the corporate customer has run out of cash (Sathye et al. 2003). The word "immediate" is used as it begs the question "Why did the corporate borrower run out of cash?" Traditionally, lenders have faced credit risk in the form of default by borrowers. To this date, credit risk remains the major concern for lenders worldwide. The more they know about the creditworthiness of a potential borrower, the greater the chance they can maximize profit, increase market share, minimize risk, and reduce the financial provision that must be made for bad loans.

Understanding and using an analytical lending framework that is based on effective analysis of the business strategy and competitive environment, accounting and financial analysis, and totally focus on the borrower's future cash flow is part of the credit culture (Palepu et al. 2000). The future or projections of cash flow is operating cash flow or its closely related metric, free cash flow, which is operating cash flow adjusted for capital expenditure (Mulford & Comiskey 2005). Borrowers' ability to generate cashflowability is an important component in any lending decision. Future cashflowability directly affects the value of loans, because it constitutes the ultimate payments expected from borrowers and hence, it is a crucial input for financial valuation models (Allen & Cote 2005). The conceptual frameworks of the Financial Accounting Standard Board (FASB) and the International Accounting Standard Board (IASB) stress the ultimate objective of financial reporting is to provide investors and creditors with quality information to predict future cash flow (Hammami 2012).

1.1 Research problem, purpose statement and contribution to knowledge

Understanding the cashflowability by commercial bankers is a very important lending target because lenders insist on a loan being repaid in cash. A good lender always understands what it takes for a business's cashflowability to be adequate to repay debt as scheduled (McGuiness 2000). The rapid generation, conservation and effective utilization of cash is the whole foundation on which a business rests. Cashflowability is a solid foundation in the various steps of credit analysis that must underlie any lending decision. One of the key skills of an effective loan officer is the ability to analyse all cashflowability techniques of a commercial borrower so as to achieve an understanding of the company's past performance and creditworthiness, and then use that analysis to make sound predictions about the company's future performance and creditworthiness (Basu & Rolfes 1995, Palepu et al. 2000).

While the cash flow statement has been a required part of the annual financial report for more the 25 years and while there has been considerable support for this statement since its proposal in 1986, as well as after the collapse of big names like Cendant, Sunbeam, HealthSouth, Tyco, WorldCom and Enron, little has been written or developed on its effectiveness use or analysis. In addition, no attempts have been made to provide a definitive study or actual use for a cash flow statement and its information and/or techniques in the lending decision. This is to say previous literature and studies were on the presentation only on the interpretation of the cash flow statement which finds that the differences of opinion among FASB members were about the classification of cash flow based on underlying transactions, for example, direct and indirect methods, and they where not about the cashflowability itself.

The main objective of the current work is to improve the lending analysis and smooth the lending decision. One way to achieve this is to examine how the lenders react to the bank's decision towards cash flow oriented and educated by focussing on and using the cash flow statement and its information and techniques as a tool in their lending decision. The study aims to empirically investigate the impact of understanding, using and analysing the borrower's cash flow statement on the commercial lending decision and exploring the "cashflowability" information and/or techniques to assess the creditworthiness of the borrowers in order to improve the lending analysis and smooth the lending decision. There are outcomes a lender can generate from the statement of cash flow which enhance the power of the cashflowability as a tool for a robust lending decision which ultimately maximizes the lender's profitability and shareholders' wealth, reduce the provision for bad loans and doubtful debts, reduce the management extra work, cost and time in attempt to recover the bad loans, smooth the borrower's operations and profitability, and reduce the borrower's cashflowability problems and costs.

This paper contributes to the literature in several ways. Previous studies examine developed countries such as the US, UK, and Australia. The current research investigates a new context of one of the growing developing countries: the Kingdom of Bahrain. Even though there is no known work that provides a study on the importance of cash flow to the commercial lending banker in the Kingdom of Bahrain, it is hoped that this research will provide a significant positive correlation demonstrating the relationship between the commercial lending decision and the cashflowability information and submit a cash flow framework that will improve credit analysis and smooth the credit decisions and management. As far as the contribution to the body of knowledge to accountancy is concerned, it is hoped that this study will establish the usefulness of cash flow statements to lenders in the Kingdom of Bahrain, add to the store of knowledge globally about the usefulness of the cash flow statement in making a commercial lending decision, and expand the scarce information in the area of cash flow lending to record whether results are consistent or not with the research hypothesis. The current study is distinct from other studies in the same field in that it addresses accounts used in the Kingdom of Bahrain and so represents the first piece of accounting research investigating the perception of the Kingdom of Bahrain's lenders (as users of financial reports) on the cash flow statement.

1.2 Research questions

The research is planned to express that a major contribution to the effective management of the credit process is the lender's thorough understanding to the borrower's cashflowability. This is because it is difficult to overstate the importance of the cashflowability to overall corporate financial health and because the fundamental concept of credit quality and valuation are based on cashflowability, particularly the operating cash flow. This research intends to replicate the previous studies in order to fill gaps from previous research, by using cash flow information and techniques from statement of cash flows. The following research inquiry is proposed:

> *"How does statement of cash flows and its information impact the commercial banks' lending decisions?"*

The major questions in this area are:

- How do commercial banks in the Kingdom of Bahrain (lenders) use the statement of cash flows and the cashflowability analysis in the credit area?
- What cashflowability techniques and information do commercial banks in the Kingdom of Bahrain use to evaluate the creditworthiness of their borrowers?
- How do the commercial sectors in the Kingdom of Bahrain (borrowers) manage their cash inadequacy (inability to pay on time) and cash insolvency (insufficient funds to meet operation needs)?

2 IDEAS AND DETAILS

Every loan booked contains a risk, that is, credit risk or default risk. The major tool a lender needs to assess the default risk is the credit analysis, which is a technique of the risk management in order to assess and verify the capacity to repay and sustainability of future cash flow must do so. While applying the credit analysis the lender must know with certainly what he/she is looking for in the financial statements, ratios and numbers, simple questions need to be answered and facts to be verified. Our idea is that lenders should be more cashflowability oriented and educated (as shown in Figure 1) to apply judgmental credit analysis.

The cashflowability conceptual framework in Figure 1 is based on the relevant literature, studies, and recommendations. Ford (1996) argues that building a cash flow framework is a critical issue for credit analysis to improve the quality and efficiency of credit decision, lead to similar credit conclusions, and it is a learning system decision for the credit analysis. Kwork (2002) suggests that accounting academic and banking professionals should put more emphasis on the value of cash flow statements and information that could be obtained and that instructing new materials on using the cash flow statement and cashflowability ratio analysis. Allen & Cote (2005) argue for the development of creditors' decision-making by understanding the cash flow information beyond the earning information. None of them had, however, examined the interpretation and actual uses of the cash flow statement for lending decision.

Our idea and judgment on Figure 1 is that since a successful bank has a credit culture that allows lending to occur without losing much, which is the fundamental conundrum of lending, an effective lender should analyse both the historical and forecast cashflowability. In addition, the historical cashflowability analysis should be focussed on the operation, investment and financing cash flow activities presented in the cash flow statement and the cashflowability techniques and/or information which are the variables or the outcomes calculated and examined from the cash flow statement. The analysis of the operating cash flow activities is to identify and quantify the primary components of operating cashflowability which should pay back the loan. The investing cash flow activities are

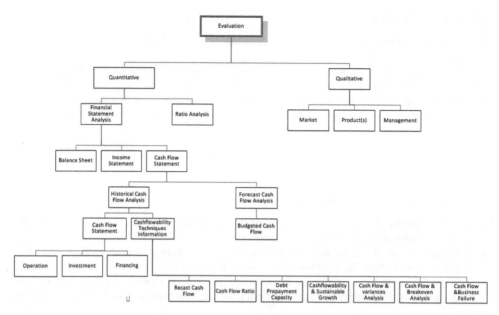

Figure 1. Cashflowability conceptual framework – analysis for lending (evaluation).

analysed to ensure that these discretionary decisions are based on sound managerial and financial strategies, whereas the financing cash flow activities are analysed to gain understanding of the business funding tactics. All of which are more accurate analyses focussing only on cash flow in and cash flow out (cashflowability), the main factors a lender investigates.

More importantly, our idea and argument is on adding the cashflowability techniques and/or information to the credit and financial analysis as they represent the central focus on the risk analysis and a way of improving the lending analysis and smoothing the lending decision (see Figure 1). By recasting or restating the cash flow statement for the lending decision purpose, the loan officer examines the free cash flow which is the closely related metric of the operating cash flow calculated by adjusting the operating cash flow for capital expenditure. While the information in a reported cash flow statement can be used to analyse the business' performance, liquidity and efficiency, it may not always be easy to do so because even though the accounting standards provide broad guidelines on the format of a cash flow statement, there is still significant valuation across firms in how cash flow data are disclosed. Therefore, to facilitate a systematic analysis and comparison across firms, credit analysts should recast the information in the cash flow statement, using their own cash flow model to examine cash remaining for creditors and shareholders. The cashflowability ratios examine the borrowers' liquidity, solvency, profitability, and efficiency on actual cashflowability activities mainly the cash flow from operation rather than accrual accounts.

To investigate how the borrowers will service the loan smoothly, the debt repayment capacity should be examined. Sustainable growth based on cashflowability should be analysed to investigate how the borrower can manage cash during the growth and how to finance the growth. To study how the cash position of the borrower is affected and how much is required due to the growth, change in profitability and change in tax efficiency, monitoring operating cash flow using variance analysis should be investigated. The cashflowability based on a cost-volume-profit model (break-even analysis) should be analysed to test how the borrower pays the principal and interest on due date. Finally, to investigate how the borrower survives and how to payback the debts, cashflowability and business failure should be examined.

The second cashflowability analysis is the forecast cash flow. Since the real risk in lending is to be found in the assessment of the repayment proposal, it is important that the source of repayment is made clear from the outset and the lender must establish the degree of certainty that the promised funds will be received. Where the source of repayment is cash flow, the lender will need projections in the form of budgeted cash flow to ensure there are surplus funds to cover repayment after meeting commitments.

3   LITERATURE REVIEW

In the United States of America the cash flow statement is prepared in accordance with the Financial Accounting Standards Board (FASB) Directive No. 95 of November 1987. In the UK, the statement is drafted in line with Financial Reporting Standard (FRS) No. 1 – Cash Flow Statement. In countries that have adopted international accounting standards, the statement is drafted in line with International Accounting Standard (IAS) No. 7 – Cash Flow Statement (Radebaugh & Gray 1997, Sutton 2000, Nobes & Parker 1998, Wild & Moon 1991). In 1985, Canada became the first country to replace its fund-flow statement by the cash flow statement. By mid-1991, the USA, New Zealand and South Africa had done the same. Britain, Australia and IASC all issued statements announcing their intention to follow suit.

Since the issuance of accounting standards on the cash flow statement, there have been studies on cashflowability reporting in various accounting fields: for example, in the USA (Gentry et al. 1990, Epstein & Pava 1992, McEnroe 1996, Kwork 2002), Australia (Jones et al. 1995, Yap 1997, Jones & Ratnatunga 1997, Jones & Widjaja 1998, Jones et al. 1998), Hong Kong (Hung et al. 1995), New Zealand (Dowds & Esslemont 1997), the UK, Japan (Koje 2012), and Qatar (Hammami 2012). All of these studies suggest the usefulness of the concept of cashflowability though they have used different sets of methodologies and sampled different respondents.

The aims of this section, in addition to reviewing and evaluating the literature dealing with cashflowability usefulness, are to review the empirical evidence concerning the cash flow statement and to discuss the key variables used by researchers and their theoretical reasons for choosing them. There is an extensive literature available, from many countries, on the usefulness of cashflowability, in particular the USA, the UK and Australia. The main studies which address the research problem and from which the research gap had been based are outlined in §3.1.

3.1   *Main studies on cash flow in lending*

Table 1 is a summary of the main studies investigating the effect of cash flow on lending decisions in date order. The studies in Table 1 illustrate the importance of the cash flow statement and cashflowability analysis for commercial lending decisions and assure that lenders' understanding of borrowers' cashflowability is essential. None of them, however, examine the interpretation and actual uses of the cash flow statement for lending decisions.

3.2   *Main studies on cash flow accounting*

Table 2 gives a summary of the main studies investigating the effect of cash flow on accounting fields in date order. It indicates that, while the studies on the information relevance of cash flow in the accounting literature provide evidence of the usefulness of the cash flow statement and its information, some studies produce conflicting results. The cash flow definition and the model inadequacy used were the main reasons for the conflicts.

3.3   *Advanced issues of cashflowability*

Cashflowability is a solid foundation in the various steps of credit analysis that must underlie any lending decision. One of the key skills of an effective loan officer is the ability to analyse

Table 1. Summary of studies investigating the effect of cash flow on lending decisions.

Study	Method	Results/Comments
Emmanuel (1988)	–	Cash flow data are helpful in interpreting the borrower's profitability, liquidity, management, and financial risk, thus enabling a lender to make an informed decision about whether to lend and how to structure the loan.
Fulmer et al. (1992)	Survey analysis of 266 bank officers to test the credit evaluation practices.	All respondents require historical financial statements for new commercial term loans to new borrowers. Ninety per cent require a statement of cash flow; 80% require a separate cash flow schedule specified by the lender. Stability of cash flow and net income over time are the most important risk factors followed by subjective factors, such as size and reputation, and key ratios.
Comiskey & Mulford (1992, 1993)	Two case examples (K. Swiss Inc. and Crown Crafts Inc.)	Cash impact summary identifies the cash flow needs caused by growth and by changes in profitability and operating efficiency. As growth stabilizes in the future, the firm's cash needs should decline. Income taxes paid cause the swing in net cash.
Eastman (1993)	–	Lenders who carefully examine prospective borrowers' cash flow statements can gain new insights into their customers' stability, quality of management and creditworthiness. Bankers should take full advantage of the information in cash statements because of the focus on cash – the only thing that repays loans.
William & William (1993)	Survey analysis of 54 banks to determine what factors influenced commercial loan approval.	Among many factors influencing the loan decision, the frequent availability of reliable and informative financial statements (balance sheet, income statement, and cash flow statement) is most important.
Ford (1996)	–	Credit analyst should focus on the uses of cash flow (cash outflows) to see how much finance is needed and why, and should focus on the sources of cash flow (cash inflows) to see the availability of cash to repay the debts and interest or repurchase the common stock and paid dividend. Building a cash flow framework is a critical issue for credit analysis. This will improve the quality and efficiency of credit decision, lead to similar credit conclusion and it is a learning system decision for the credit analyst.
Ahadiate et al. (2002)	A cross-cultural study of American and Asian-Pacific bankers.	All commercial bank lenders require a balance sheet and income statement as a basis for lending decision, and very few of those lenders require the cash flow statement. Asian companies in their home countries apply the cash basis of accounting. In the US, however, they use the accrual-basis of accounting which complies more with the US GAAP.
Kwork (2002)	Verbal-protocol analysis on loan officers, financial analysts, academics, and auditors, to investigate the effect of cash flow statement format on lenders' decisions.	Absence of the cash flow statement will not change the lending decision since interested lenders can obtain information about the cash flow from the balance sheet and notes to the accounts. Lenders usually ignore the information provided in the statement of cash flow. Most lenders do not refer to the cash flow situation because they are uncomfortable with information in this report. They also do not refer to the notes related to the SCF under the direct method. The main reasons for relying on balance sheet information are that cash flow statement is considered a new statement in comparison to the balance sheet and lack of training in utilizing the SCF. New materials on using the SCF and SCF ratio analysis need to be instructed by accounting academic and banking professionals. The accounting standards setters should mandate the direct method to the SCF only.

Table 1. Continued.

Study	Method	Results/Comments
Allen & Cote (2005)	Experimental Study.	Similarity between creditors and investors in analysing the accrual earnings over the operating cash flow in predicting the organizations' performances. Lenders' practical perception is similar to general theoretical perception on favouring the earnings over operating cash flow.
Koje (2012)	Experimental study.	Extends the work of Klammer & Reed (1990) but examines accounting students rather than loan officers and finds that direct method of cash flow statement was preferred by account users over indirect method for decision making. His result was within both FASB and IASB.

all cashflowability techniques (basic and advanced) of a commercial borrower so as to achieve an understanding of the company's past performance and creditworthiness, and then to use that analysis to make sound predictions about the company's future performance and creditworthiness. A prudent loan officer must be able to determine which of the advanced techniques presented in this section are appropriate in a given situation.

Very few studies have examined the borrower's debt repayment capacity particularly its association with the cashflowability (for example, Ellison & Lane 2003, Rizzi 1994, Leibowitz et al. 1990). According to Leibowitz et al. (1990) and Milling (2000) how much leverage a firm can support is not an easy question a commercial bank lender can answer. Both argue that healthy and stable cashflowability can attract the lender since predictable cash flow offers a high probability that debt payment obligations will be met. Leibowitz et al. (1990) argue that creditors at a certain level of cash flow know exactly the amount of cash to be supported. They further argue that creditors must take into consideration the cashflowability volatility because borrowings are only allowed with a high probability of sufficient cash to repay the obligations. That is to say, a prudent lender should always relate the amount of debt he/she would be willing to supply to the level and riskiness of the borrower's cashflowability and particularly the amount of net cash flow from operation. Today, some effective lenders use an easier and more sophisticated method to determine debt capacity. The cashflowability method is now used for highly leveraged transactions (Rizzi 1994) because those lenders are particularly concerned with cash flow to pay debt (Ellison & Lane 2003).

No effective studies have been made on the sustainable growth usefulness for lenders. The only articles found are by Eiseman (1982) and Kester (2002). According to Kester (2002) the sustainable growth rate can be a valuable tool for commercial lenders in assessing a borrower's ability to plan for and manage growth in a financially sound manner. A professional lender knows that for growing companies there is a growth rate; this rate is the rate at which they can grow, without increasing leverage providing that they maintain the dividend payout ratios, after tax margins, net working assets activity ratios, and the sales/fixed assets ratio. Some effective lenders nowadays have developed a sustainable growth rate based on cashflowability rather than accrual accounts.

Researchers who examined the role of operating cash flow in predicting the probability of default, however, are very few; they include Casey & Bartczak (1985), Gentry et al. (1987), Bernard & Stober (1989), Aziz et al. (1988), Gilbert et al. (1990), Ward (1994) and Charitou et al. (2004). All of these researchers agreed that the failure of a business was mainly due to the lack of cash from operations, which leads to inability to serve the debts. More importantly, all the above studies with the exception of the first study by Casey & Bartczak (1985) admit the usefulness of cash in predicting the business failure. The study of Aziz et al. (1988) on the cash flow reporting and financial distress proves the power of both cashflowability based and mixed (cashflowability based, Z and Zeta) over the Z and Zeta in predicting the corporate bankruptcies. It also proves that the previous studies of Casey & Bartczak (1985) and Gentry et al. (1987), which favoured the Z and the Zeta models are retrogressive steps.

Table 2. Summary of studies investigating the effect of cash flow on accounting field.

Study	Method	Results/Comments
Lee (1982)	A survey to 488 members of the Institute of Chartered Accountants to test the importance of cash flow accounting (CFA).	CFA was seen as useful to many of the main financial report user groups. There were no particular problems in obtaining cash flow data for reporting purposes. Forecasted CFA was seen as a more limited system.
Gentry et al. (1990)	Annual financial statement data for a sample of 333 companies to calculate standardized value for 13 cash flow components over the 1982–1986 period.	Cash flow components as percentage of total cash flow vary across company size and industry group. These cash flow profiles can serve as a reference or benchmark for comparative analysis. Encourage financial analysts to use cash flow analysis.
Unknown author	Cox Proportional hazard model and bank cash flow information to determine whether adding cash information will improve current bank failure prediction method.	While banks uniquely have access to an insured source of funds (i.e., deposits), they still need to generate positive cash inflows and control cash flow if they are to remain financially strong and competitive. Cash flow information should add to the knowledge of banks' actual financial conditions just as it does for non-bank firms. As banks move more and more to off-balance sheet activities, the traditional balance sheet ratios become less and less able to reflect the banks' condition. The banks' ability to generate measurable, positive cash flows becomes a more important indication of bank health.
McEnroe (1996)	A survey analysis to examine the attitude of accounting professions, accountants (both public and private), financial analysts, and investment advisors towards additional cash flow disclosures as well as their perceptions regarding certain professed attributes of cash accounting (CFA).	Financial analyst and investment advisors were significantly more receptive to CFA as integral part of the external financial reporting framework than accounting professors and accountants, because the latter groups have more exposure to US GAAP. Financial Analysts and investment advisers, as opposed to the other two groups, were favourably disposed to the requirements of per share of cash flow from operating activities. Investment advisers were also in favour of cash flow forecasts. Financial analysts demonstrated the most support for a free cash flow statement, while remaining groups mainly expressed uncertainty. Respondents did not express an overwhelming preference for the direct versus the indirect methods. Therefore the findings support the argument of flexibility in the format of the operating cash flow section of the cash flow statement.
Yap (1997)	A survey to provide an understanding of the demand for cash flow statement (CFS) by users and their perceptions of the usefulness of conventional financial statements.	CFS is regarded as an important source of information in financial decision making. It is used by a majority of respondents, particularly for the evaluation of liquidity, solvency and financial flexibility. While cash flow and liquidity were ranked second after future prospects and level of profit (66% nominated future prospects as extremely important, 62% level of profits, and 55% nominated cash flow), 70% of respondents indicated that CFS had improved the usefulness of reports.

Table 2. Continued.

Study	Method	Results/Comments
Jones & Ratnatunga (1997)	Hypothesis study to address the decision usefulness functions of the cash flow statement (CFS) in the context of three specific variables: firm size; financial reporting focus to users; and voluntary CFS preparing practice.	Each of the hypotheses based on these variables have been justified as a relevant test of the decision usefulness of CFS. CFS has relevance across a number of internal and external decision making contexts.
Wallance et al. (1999)	A study of the characteristics and comprehensiveness of disclosure in cash flow statements (CFS) published in the 1995 annual reports of UK firms and their relationships with selected firm-specific characteristics.	Cash flow reporting comprehensiveness is an increasing function of firm size while that comprehensiveness is decreasing function of return sales. Manufacturing and trading firms and negative net cash flow firms tend to release more comprehensive CFSs than service and conglomerate firms and positive net cash flow firms.
Unknown author	An accounting research to examine the articulation in cash flow statements by using actual corporate financial statements.	Infrequent occurrence of a clean articulation of operating activities in the cash flow statement under the indirect method even when line items are examined one at a time. The more balance sheet and income statement items included in the analysis the less clean articulation will be. Most of investing and some of financing activities items in the CFS are not easy to investigate because of different methods and ways of reporting them, e.g., securities, plant assets, debts (short-term, long-term, leases, etc.) They usually involved an explanation within the notes and a cross line comparison is not useful. Therefore, investment and financing items show lower clean articulation among selected firms. The larger the firm (e.g., higher total assets) the more complex the reporting and the less clean articulation it has.
Unknown author	The relationship between brand strength, revenue, operating cash flow and shareholder growth (1998–2000).	While a strong relationship emerged between brand strength and growth of sales and cash flow, the correlation with shareholder returns (share price growth plus dividends) was particularly notable.
Charitou et al. (2004)	Neural network and logit methodology to a dataset of 51 matched pairs of failed and non-failed UK public industrial firms over the period 1988–1997.	A parsimonious model that includes three financial variables: cash flow, profitability and financial leverage variables yield an overall correct classification accuracy of 83% one year prior to failure. Ability of the Altman model (Z-score) did not perform that well compared to other models tested and thus it may not be appropriate for UK business failure. Operating cash flow process discriminating power when it comes to predicting UK company failure (contrary to prior studies).

No effective studies on cashflowability variance were found by the authors. The only two related articles found are by Hull (1990) and Comiskey & Mulford (1992). Without reference to the fundamental causes for the cashflowability drain in the short term, the lender might make an incorrect assessment of the firm's future debt-servicing capacity, dividend payments, and meeting its short-term obligations (Comiskey et al. 1991, Hull 1990).

To make an accurate cashflowability analysis, however, accrual-based profitability must be examined together with cashflowability before assessing the true health of a company (Beach 1985). According to Beach, an agreement of both cashflowability and profitability clearly indicates either a profitable company with a positive cash flow or a failing company experiencing a cash drain, while a disagreement of the two elements required closer scrutiny of a company in order assess its financial health accurately. Although, as discussed earlier, a traditional analysis coupled with a careful cash flow statement analysis provide a great deal of useful information, an alert commercial bank loan officer can take a further step to make a cashflowability variance analysis. The only comprehensive study on cashflowability break-even is found by Jerris & Tennant (1991).

Cashflowability analysis helps the lending officer to be more able to assess the continuances of the surplus or shortage of the cash flow (Comiskey & Mulford 1992) and hence understand the financial operations of the borrowers before reaching a healthy commercial lending decision. The availability of literature concerning the presentation of the cash flow statement and the lack of articles concerning the interpretation of the cashflowability, have been observed in the preceding discussion. According to Weston & Brigham (1990) there is not yet a generally accepted systematic framework widely used to analyse cash flow statements such as those found for analysing income statements or balance sheets. A few authors have, however, suggested frameworks for analysing the cashflowability such as Brian (1993), Carslaw & Mills (1991), Figlewiez & Zeller (1991), Henderson & Maness (1989), Siegel & Akel (1989) and Gahlon & Vigeland (1988). According to Brian (1993) only little insight could be gained of the operating cost structure from the GAAP-prepared financial statements; however, the cash flow statement can be used to gain that insight.

4 METHODOLOGY

4.1 *Research strategy*

This deductive study will explore the relationship between the commercial bank lending decision (dependent variable) and the demand for cash flow information (independent variable). Based on a survey of commercial banks engaged in corporate lending in the Kingdom of Bahrain it is hoped to report on the degree to which Bahraini commercial banks incorporate cash flow considerations into their corporate lending decisions. The aim of the survey is to collect opinion and behaviour of the commercial banks in the Kingdom of Bahrain on cash flow. Based on the research hypotheses, this study will focus on two major variables: the lending decision and statement of cash flows. In the prediction model, the lending decision will be investigated as a dependent variable caused by independent variables, that is, the statement of cash flows.

4.2 *Data collection methods and material*

A quantitative method based on postal and online questionnaires to commercial banks' loan officers and credit analysts in the Kingdom of Bahrain will be used (the research sample). This will be short and close-ended. It is anticipated that the principal focus will be to elicit respondents' beliefs about the importance of statements of cash flow to the lending decision. This research also investigates the ability of cash flow information and techniques for a robust lending decision. Even though there is no empirical evidence supporting this information and the techniques except the cashflowability and bankruptcy prediction, this study initiated them as a sign of good and sophisticated lending decisions. Eight cash flow information and techniques will be tested: recast cash flow, forecasted or budgeted cash flow, cash flow ratios, debt repayment capacity, cashflowability and sustainable growth, cashflowability and bankruptcy prediction, monitoring operating cash flow using variance analysis, and cashflowability based cost-volume-profit model (break even analysis).

4.3 Data analysis method

The above information and techniques will be evaluated using data from statements of cash flow. This research will utilize quantitative methods in which the data are analysed based on statistical techniques, which include descriptive statistics, correlation, and regression analysis. The descriptive statistics provide an initial summary data of the essential features of the sample. The correlation analysis is used to fundamentally examine the relationship between dependent and independent variables. Regression analysis is applied to test the prediction models depending upon the ability of cash flow outcomes (information and techniques) to impact lending decision. All analytical techniques use the computer software package Statistical Processing for Social Scientists (SPSS).

5 CONCLUSIONS

The aim of this paper has been to contribute to the development of a general, systematic, and analytical framework so that the cash flow statement can fulfil its potential. The major tool a lender needs to assess the default risk is the credit analysis, which is a technique of the risk management, in order to assess and verify the capacity to repay, and the sustainability of future cash flow to do so. While applying the credit analysis the lender must know with certainly what he/she is looking for in the financial statement, ratios and numbers, simple questions need to be answered and facts to be verified. In our opinion, the credit analysis should be more cashflowability orientated and both loan officers and credit analysts should be cashflowability educated and build their technical skills in accounting, finance, economics, risk management, and human skills in management and psychology (for example, marketing and communication). This is because when they can apply these disciplines, they are ready to apply judgmental credit analysis based on cashflowability, the only thing that pays back the loan smoothly and on time. It is worth mentioning that each cashflowability technique in this paper is a field which deserves future comprehensive academic studies and investigations, particularly by employing non-American data sets since the majority of the cashflowability studies are American.

REFERENCES

Ahadiat, N., Pak, H. & Salimi, A. 2002. Financial accounting in commercial lending institutions: A cross-cultural study. *Journal of Commercial Bank Lending* (2002) pp. 101–112.

Allen, M.F. & Cote, J. 2005. Creditors' use of operating cash flows: An experimental study. *Journal of Managerial Issues* 17(2): 1–8.

Aziz, A., Emanuel, D.C. & Lowson, G.H. 1988. Bankruptcy prediction: An investigation of cash flow based models. *Journal of Management Studies* 25(2): 419–437.

Basu, S.N. & Rolfes, H.L. (Jr) 1995. *Strategic Credit Management*. NY: John Wiley & Son.

Beach, R. 1985. Cash flow *vs* cash flow. *Commercial Lending Reviews* 86(48): 48–52.

Berned, V.I. & Stober, T.I. 1989. The nature and amount of information in cash flow and accruals. *Accounting Review* 64(4): 624–652.

Carslaw, C. & Mills, J.R. 1991. Developing ratios for effective cash flow statement analysis. *Journal of Accountancy* 172(5): 63–70.

Casey, C.J. & Bartczak, N.J. 1985. Using operation cash flow data to predict financial distress: Some extension *Journal of Accounting Research* 23(1): 384–401.

Charitou, A., Neophytous, E. & Charalambous, C. 2004. Predicting corporate failure: Empirical evidence for the UK. *European Accounting Review* 13(3): 1–34.

Comiskey, E., Gulbrandsen, S. & Mulford, C. 1991. Improving the accuracy of computer-generated cash-flow statements. *Commercial Lending Review* 6(3): 11–27.

Comiskey, E.E. & Mulford, C.W. 1992. Finding the cause of changes in cash flow. *Commercial Lending Review* 7(3): 21–40.

Comiskey, E.E. & Mulford, C.W. 1993. Understanding the reasons behind changes in cash flow. *Commercial Lending Reivew* 8(1): 29–43.

Dowds, J. & Esslemont, D. 1997. The usefulness of the statement of cash flows: Evidence from New Zealand analysts. *Accounting Forum* 21(2): 244–246.

Eastman, K. 1993. Credit analysis: Gaining insight from the statement of cash flows. *Commercial Lending Review* 8(2): 63–68.

Eiseman, P.C. 1982. Another look at sustainable growth. *Journal of Commercial Lending* 67(2).

Ellison, L. & Lane, L. 2003. Financial leverage through risk management: Leverage and the lender. *Farm Credit Canada* 14: 125–128.

Emmanuel, C.B. 1988. Cashflow reporting, Part 2: Importance of cash data in credit analysis. *Journal of Commercial Bank Lending* 70(10).

Epstein, M.J. & Pava, M.L. 1992. How useful is the statement of Cash flows? *Management Accounting (US)*, 74(1): 52–56.

Figlewicz, R.E. & Zeller, T.L. 1991. An analysis of performance: Liquidity, coverage and capital ratios from the statement of cash flow. *Ahron Business and Economic Review* 22(1): 64–81.

Ford, J.K. 1996. Credit analysis: A cashflow framework for credit analysis. *Commercial Lending Review* 11(4): 104–110.

Fulmer, J.G., Gavin, T.A. & Bertin, W.J. 1992. What factors influence the lending decision?: A survey of commercial loan officers. *Commercial Lending Review* 17: 64–70.

Gahlon, J. A. & Vigeland, R.L. 1988. Early warning signs of bankruptcy using cash flow Analysis. *Journal of Commercial Bank Lending*.

Gentry, J.A., Newbold, P. & Whitford, D.P. 1987. Funds, flow components, financial ratios and bankruptcy. *Journal of Business of Finance and Accounting* 14(4): 595–606.

Gentry, J.A., Newbold, P. & Whitford, D.P. 1990. Profiles of cash flow components. *Financial Analyst Journal* 46(4): 41–48.

Gilbert, L.R., Menon, K. & Schwartz, K.B. 1990. Predicting bankruptcy for firms in financial distress. *Journal of Business Finance* 17(1): 161:171.

Hammami, H. 2012. The use of reported cash flows *versus* earnings to predict cash flows: Preliminary evidence from Qatar. *Business Systems Review* 1(1): 103–121.

Henderson, J.W. & Maness, T.S. 1989. *The Financial Analyst's Deskbook: A Cash Flow Approach to Liquidity*. New York, NY: Van Nostrand Reinhold.

Hull, J. 1990. Monitoring a company's operating cash flow using variance analysis. *Accounting Horizon*: 50–57.

Hung, H.Y., Chan, M. & Yiu, A. 1995. The usefulness of cash flow statements. *Asian Review of Accounting* 3(1): 192–104.

Jerris, S.I. & Tennant, K. 1991. A simplified cash-flow-based. cost-volume-profit model for use in lending decisions. *Journal of Bank Cost & Management Accounting* 4(2): 77–89.

Jones, S. & Ratnatunga, J.K. 1997. The decision usefulness of cashflow statements by Australian reporting entities: Some further evidence. *British Accounting Review* 29(1): 67–85.

Jones, S., Romano, C. & Smyrnios, K. 1995. An evaluation of the decision usefulness of cashflow statements by Australian report entities. *Accounting and Business Research* 25(98): 115–129.

Jones, S., Sharma, R. & Mock, K. Managerial evaluations of the relevance of cash *versus* accrual-based financial reports in the Australian food industry. *Australian Accounting Review* 8(16): 51–58.

Jones, S. & Widjaja, L. 1998. The decision relevance of cash-flow information: A note. *Abacus*, 34(2): 204–219.

Kester, G.W. 2002. How much growth can borrowers sustain? *RMA Journal* 84(10): 49–53.

Klammer, T.P. & Reed, A.A. 1990. Operating cash flow formats: Does format influence decisions? *Journal of Accounting and Public Policy* 9: 217–235.

Koje, K. 2012. Decision usefulness of cash flow information format: An experimental study. *International Review of Business* 12(3): 23–44.

Kwork, H. 2002. The effect of cash flow statement format on lenders' decisions. *International Journal of Accounting* 37(3): 347–362.

Lee, T.A. 1982. Laker Airways: The cashflow truth. *Accountancy* 93(1066): 115.

Leibowitz, M., Kogelman, S. & Lindenberg, E.B. 1990. A shortfall approach to the creditor's decision: How much leverage can a firm support? *Financial Analysts Journal* 46(3): 43–52.

McEnroe, J.E. 1996. An examination of attitudes involving cash flow accounting: Implications for the content of cash flow statements. *The International Journal of Accounting* 13(2): 161–174.

McGuiness, Bill. 2000. *Cash Rules*. Kiplinger Washington Edition, Inc.

Milling, B.E. 2000. *Cash Flow Problem Solver: Common Problems and Practical Solutions*. Chilton Book Company.

Mulford, C.W. & Comiskey, E.E. 2005. *Creative Cash Flow Reporting: Uncovering Sustainable Financial Performance*. New York, NY: John Wiley & Sons.

Nobes, C. & Parker, R. 1995. *Comparative International Accounting* (fifth edition). Prentice-Hall Europe.

Ohlson, J.A. 1990. Financial Ratios and the Probabilistic Prediction of Bankruptcy, 18(1): 109–113.

Palepu, K.G., Healy, P.M. & Bernard, V.L. 2000. *Business Analysis and Valuation Using Financial Statements*. South-Western College Publishing.

Radebaugh, H.L. & Gray, J.S. 1997. *International Accounting and Multinational Enterprises*, fourth edition. New York, NY: John Wiley & Sons.

Rizzi, J. 1994. Guaging debt capacity. *Corporate Cashflow* 15(2): 33–37.

Sathye, M., Bartle, J., Vincent, M. & Buffey, R. 2003. *Credit Analysis and Lending Management*. John Wiley & Sons Australia.

Siegel, J.G. & Akel, A. 1989. A financial analysis and evaluation of the statement of cash flows. *The Practical Accountant*.

Sutton, T. 2000. *Corporate Financial Accounting and Reporting*. Englewood Cliffs, NJ: Prentice-Hall.

Wallance, R.S., Olysegun, Choudhury, M.S.I. & Adhikari, A. 1999. The comprehensiveness of cash flow reporting in the United Kingdom: Some characteristics and firm-specific determinants. *International Journal of Accounting* 34(3).

Ward, T.J. 1994. Cash flow information and the prediction of financially distressed mining, oil and gas firms: A comparative study. *Journal of Applied Business Research* 10(3): 78–86.

Weston, F.J. & Brigham, E.F. 1990. *Essentials of Managerial Finance*. Orlando, Frorida: Dryden Press (ninth edition).

Wild, K. & Moon, J. 1991. *Cashflow Statement – A Practical Guide*. Touche Ross.

William, P.J. & William, C.K. 1993. Credit analysis: Survey shows lenders' precepts constant over two decades. *Commercial Lending Review* 88–102.

Yap, C. 1997. Users' perceptions of the need for cash flow statements: Australian Evidence. *European Accounting Review* 6(4): 653–672.

Quotas and empowerment: Enhancing female leadership in corporate boardrooms

Layla F. Alhalawachi & Samia Costandi
Ahlia University, Kingdom of Bahrain

ABSTRACT: This paper focusses on understanding three important concepts namely, empowerment, female leadership, and quotas regarding female representation in the corporate boardroom within the private sector. Through a review of the research, the aim is to draw a landscape of existing literature on the topic of quotas for women. This paper explores whether the representation of women in the corporate boardroom could be enhanced through applying the concept of mandatory quotas across different industries. The study uses the current status of women within the boardroom in the private sector as its context and builds upon contemporary knowledge in the field. By linking the concept of quotas to female empowerment and leadership in the corporate boardroom, the paper hopes to fill a theoretical gap that has existed in the literature regarding whether mandatory quotas would be effective in eliminating discrimination against women in the boardroom or not.

1 INTRODUCTION

Although it is self-evident that women have since the dawn of history contributed immensely to the economic growth of countries, the issue of the extent of their contributions remains controversial. Women have contributed through their bearing and raising of children, taking care of homes and families, cooking and doing domestic work, nurturing and teaching their young, taking care of the elderly in their families, gathering and hunting alongside males in early societies, collecting fuel, taking care of animals, planting and growing vegetables, doing laundry, taking part in agriculture and industry in modern societies; despite all that, according to (Burn 2005) "their important economic contributions have been frequently forgotten or devalued as 'natural'". Women have not been officially recognized as contributing directly and visibly to the economy up to this day because most have been involved in unpaid labour which is not factored into GDPs or economic indexes. While women have come a long way and have legitimately struggled and procured a substantial part of their usurped human, civil and legal rights, and despite the fact that they have demonstrated that they have been instrumental in the development of global societies and the rise of civilizations, once "societies based on money evolved, men's labor appeared to have greater value because it was done for money or the exchange of goods" (Burn 2005). Hence, the issue of women's contribution to the labour force and the economy of countries remain debatable.

Since married women are usually expected to bear the primary responsibility for housework and child care, it is not surprising that marriage and children reduce the likelihood that women will work for pay. When such women do work in the business market, they generally assume the "double burden" of working at a paid job while maintaining nearly complete responsibility for home work (Ruth 1990). In recent years women all around the world have experienced an exponential increase in participation in economic activities outside the home. Such an increase in participation is evident from women not only being able to join the labour force but also taking up positions in various areas that were previously only considered the domain for men (Goodman et al. 2003). Women have faced many challenges in their attempts through feminist movements to break those barriers and gain equal opportunity for employment within society (Ruth 1990) and there is a dearth of research in the area which has caused difficulty for both business organizations and women. Through an analysis of the relevant literature, this paper hopes to draw out the essential aspects that impact women's representation in the boardroom. There is a need to gain new insights into how to enhance female representation at important organizational levels: this paper hopes to do that.

2 BACKGROUND OF THE STUDY

Female representation in occupations in general is a well-researched area that has attracted the attention of researchers (e.g. Moghadam 2003). The literature review related to the participation of women in the labour force has identified a number of concepts that could be effectively tackled to enhance the positions occupied by women in various organizations. For instance, empowerment and leadership are two key concepts discussed by many researchers (Quinn & Spreitzer 1997) as affecting women's employment status and participation in the labour force of organizations.

In the same vein, another important concept that has attracted the attention of researchers is organizational settings (Goodman et al. 2003). Researchers have found that the boardroom specifically is an organizational setting where there is a lack of female representation in the corporate sector (Singh & Vinnicombe 2004). Boardrooms have been generally occupied by men, and there is a perceptible lack of representation of women on those boards in significant numbers. The literature shows that there are factors that could be contributing to this anomalous situation, and such factors include lack of empowerment of women (OECD 2012), [what is perceived as a] lack of leadership attributes among women (Eagly et al. 2003), and lack of quota (Rai 2012).

There are significant shortcomings at the conceptual level and a lack of in-depth analysis in the identification of those factors affecting female representation in the boardroom. There is a need to single out and sift the factors that limit the empowerment of women and impact leadership. This paper has chosen three important concepts for investigation, the outcome of which is expected to contribute to the contemporary body of knowledge. The factors are: empowerment, leadership, and quotas in the boardroom.

The interplay between the three above-mentioned concepts is another area that has been neglected by the research community. In order to understand the above-mentioned concerns, a comprehensive literature review is offered next. This paper limits the investigation to only three concepts at this stage because including more concepts would require a more complex investigation and analysis which do not fall within the scope of this research.

3 LITERATURE REVIEW

The first and most obvious question to pose in this discussion is "what is empowerment?" An in-depth look reveals that there are not only multiple forms of empowerment, but there are also multiple definitions and multiple variables which affect perceptions of empowerment. To break this simple question of "what is empowerment?" into its constituent parts, it is sensible to consider each of these aspects in turn. Therefore this segment of the discussion will relate to (i) definitions of empowerment; (ii) forms of empowerment; and (iii) the variables which affect perceptions of empowerment. It is posited that, having critically analysed these three fundamental tenets of empowerment, it will be possible to move on and examine these aspects within the context of female empowerment. The aim at this juncture is to understand why empowerment is a useful term to explore when analysing the position of women in the labour market.

3.1 *Definitions of empowerment*

The term "empowerment" is broad and multi-faceted, embracing a number of concepts relative to the perception of the individual(s) and the contextual setting of the discussion. It is observed in the literature that the vast majority of scholars believe that the term is still not appropriately defined

(Cox et al. 2006, Denham-Lincoln 2002, Pastor 1996, Cook 1994). Burn (2005) quotes Datta & Kornberg (2002) when she defines empowerment as follows:

> "Empowerment includes the processes by which women gain greater power over their own lives both within and outside the home, and their power to bring about change in situations of gender inequality…a commitment to breaking down the structures that keep women lower in status and power."

Having said that, there is also agreement at a universal level that there are significant variations in understanding the term and it is necessary to interpret empowerment with respect to the fine-tuned details of pragmatic application (Collins 1994, Lee & Koh 2001, Denham-Lincoln et al. 2002, Shapira et al. 2010). Denham-Lincoln et al. (2002) suggest that, in its most basic form, empowerment can be defined as "a humanistic device to improve the quality of working life for ordinary employees"; however, critics such as Collins (1994, p. 18) argue that it is frequently misinterpreted as the latest management "fad" applied to shift risk and responsibility from a top-down perspective. Cooney (2004, p. 677) argues that despite on-going debates in the field of empowerment "there is no settled idea of what it actually is".

In the range of female empowerment and the notion of the "poisoned chalice" discussed by Ryan & Haslam (2004), a female in a position of power within an organization can have notional power because of her role, but limited actual power because she is unable to change others' perceptions of her capabilities. In this instance it is argued that females are not in fact empowered because at best they have pyrrhic power which is of virtually no use, making them a figurehead in the organization. Ultimately, and in relation to quotas, such an understanding of empowerment means simply giving women a position on the board to balance gender quotas, which in effect could be more damaging than useful! Authentic empowerment, on the other hand, involves the more subtle understanding of being able to influence others.

3.1.1 *Forms of empowerment*

Broad reading in this area indicates that there are approximately 15 alternative factors which contribute towards an holistic definition of empowerment. The terminology "approximately" is used because of the nuances regarding the divisions and distinctions between empowering factors. These range from the obvious, as noted by Rodrigues (1994), which are about the power to make decisions through the subtle ability of someone to influence others' perceptions via their role or task (Whiteside et al. 2011).

The implication of this in organizational contexts is that there can be an unspoken disconnection between the role and the individual. In practice this means that a role attracts power but even an individual without the status of the role is equally capable of being empowered if they command respect and have leadership capabilities giving them a horizontal influence of power and not a vertical one (Northouse 2010). Arguably, this is more likely to occur in individualist cultures where non-hierarchical organizations are more common; on the other hand, are typically accrued experience, and this can also lead to forms of empowerment through knowledge garnered from experience (Kumra & Vinnicombe 2010). This gives rise to the suggestion that there are two types of empowerment, which for the purpose of this discussion have been nominally titled (i) true empowerment, and (ii) notional empowerment. The distinction between the two has been developed after a review of the perspectives of empowerment. The conclusion has been reached that it is possible to have notional empowerment in the form of a role which should hypothetically confer power but does not do so against actual empowerment relating to the ability to influence others.

3.1.2 *Variables which influence perceptions of empowerment*

It has emerged from the literature that there are a number of variables or factors which influence perceptions of empowerment. This line of reasoning builds upon a distinction between true empowerment and notional empowerment, and seeks to explain how a divergence between the two has arisen. An example of this may be how it is possible for an individual to hold a role which should accrue a position of power and yet this individual is broadly ignored. It is argued that understanding this distinction is critical because this will explain how and why females may or may not be empowered within a role, and will also serve to explain why it is that simply giving females a nominal or notional role in a senior executive position to simply fulfil a quota may in fact do more harm than good. Effectively this work seeks to describe how the scenario of the glass cliff is generated within an organization.

3.1.3 The glass cliff

The concept of the glass cliff was first published by Ryan & Haslam (2005a,b; 2007a,b). In essence, they proposed a theory which is an extension of the glass ceiling and refers to the situation as alluded to by Peeters (1991), whereby women who are able to reach senior executive positions in organizations have in fact been set up to fail. In essence others within the organization contrive to place them in a precarious position within the business and task them with implementing a strategic change or other major project which is all but impossible.

Subsequent research by (Bruckmüller & Branscombe 2010) refers to the situation as being similar to giving female leaders in high-profile positions the "poisoned chalice", the analogy having been derived from English folklore whereby an individual is given an apparently valuable gift which later transpires to be more harmful than beneficial. In this scenario the chalice is of course the much desired high-profile leadership position which many women are unable to attain, but the poison is the fact that the work they are tasked with undertaking is all but impossible to complete (Brescoll et al. 2010; Haslam & Ryan 2008).

Although the glass cliff concept was widely acclaimed after its first publication, subsequent reflection has led to a number of criticisms of this concept. Ryan & Haslam (2004) provide considerable evidence which illustrates that "scaling the glass cliff", as they put it, is decidedly more difficult for females than males; however, Ryan & Haslam (2004) may have over-emphasized this challenge. They point out that there are sufficient females at a senior corporate level to illustrate that women can indeed reach the boardroom should they so wish.

3.1.4 The glass ceiling

The concept of the "glass ceiling" was formally identified as far back as 1995. Described by the Federal Glass Ceiling Commission (FGCC), it is the "…unseen, yet unbreachable barrier that keeps minorities and women from rising to the upper rungs of the corporate ladder, regardless of their qualifications or achievements" (FGCC 1995).

Although several early research projects looking into the existence of the glass ceiling specifically focussed on the role of women in the workplace, wider research rapidly established that the concept could be equally applied to other minority groups (McDowell et al. 1999, Davies-Netzley 1998, FGCC 1995). Research by (Cotter et al. 2001) determined that four factors always presented themselves where the glass ceiling was found to exist. These factors can be summarized as follows:

1. The existence of "a gender or racial difference" that could not be logically explained with reference to other job requirements or characteristics.
2. A statistically disproportionate difference in the presentation of the requisite minority group. For example, the gender distribution of the population is broadly 50:50; statistically, however, female representation in the boardroom even in a democratically advanced nation is at best represented by the ratio men:women = 90:10.
3. A statistically disproportionate representation which increases exponentially the more senior the role in the organization
4. The degree of inequality increases over the course of an individual's career.

3.1.5 Empowerment in the field of business and management

A great deal of literature addresses the topic of empowerment in an organizational context. Academics and practitioners concur that organizations which are empowered are more successful and have more productive employees who are engaged with the organization and, therefore, devote more of their personal resources to achieving organizational aims (Hu et al. 2012, Cox et al. 2006,

Purcell & Hutchinson 2003). The positioning of empowerment within the field of organizational studies and HRM links the concept closely with organizational and leadership success.

3.2 Leadership and empowerment

3.2.1 The relationship between leadership style and empowerment

It has been suggested by Hinkin (1995) that true leaders empower themselves to empower others on the basis that leadership is the art and science of motivating others to follow and to contribute to the overall success of an endeavour. The debate in respect to whether leaders are born or made will doubtless continue as it has yet to reach any satisfactory conclusion. There is broad consensus amongst scholars, however, that "successful" or "good" leaders (the terms being relative) possess certain traits which empower themselves and their followers (Guillaume & Telle 2011, Grey 1999, Willmott 1984). The consensus, such as it exists, holds that, when leaders, and thus their employees, are empowered, an organization performs more effectively because employees align themselves with the aims and objectives of the leader (and by proxy the organization), with the result being that collective effort and energy is poured into the achievement of these aims. The concentrated focus on achieving what is considered best at a collective level is regarded by Yammarino et al. (1993) as an example of empowered leadership. They posit that because effort and energy are centred on positive activity, as opposed to negative reactive behaviour, the result is an organizational output which is greater than the sum of its parts. For example, employees who do not feel that they must perpetually justify themselves to their leader can invest their efforts and energy in proactive tasks, as opposed to reactive tasks which are not value-adding. According to Stanley (1994) and Gajjala et al. (2010) this can only occur when there is mutual trust and respect, which is the consequence of an empowered organization managed by an empowered leader.

Empowerment is clearly acknowledged as a route to enlightened leadership and successful organizational management; thus, the logical conclusion seems to be that organizations that wish to do well would be prudent to adopt empowered leadership approaches and styles. As the preceding evidence demonstrates, however, this does not seem to be the case and many organizations appear to perform sub-optimally because of their prevailing leadership style and organizational culture (Fleming & Spicer 2003, 2007). There is a body of research which focusses on the correlation between leadership and organizational culture. The prevalent belief is that the two are to a large extent overlapping and self-reinforcing, especially where an organizational leader is particularly charismatic (Hales 2001, High-Pippert & Comer 1998, Gajjala et al. 2010). This dimension and its wider effects will be explored in greater depth at a subsequent point as extrinsic and intrinsic variables which impinge on empowered leadership.

3.2.2 Leadership style and organizational culture

There is a natural link from discussions surrounding leadership and empowerment to the correlation between leadership style and empowerment. Quite understandably different leadership styles lend themselves more readily to an empowered approach than others (Smircich 1983, Hinkin & Schriesheim 1989, Knudsen & Wærness 2009) and thus it is useful to consider which clearly identifiable leadership styles are more useful in this context. This is an area of knowledge which has also been subjected to considerable debate over the last thirty years (1985–2015) since there have been substantial changes in employee expectations culturally and socially. The significance of this is that where the debate has moved on, there has been an increased preference for an "empowered" leadership style.

The increased interest in empowered leadership styles in a Western context has, however, highlighted a growing divergence from leadership styles as favoured in other cultural settings (Abbot et al. 2005). This is evident in cross-cultural leadership situations which can become challenging to manage when fundamental differences relating to deeply-embedded cultural preferences and perceived leadership styles are in conflict with one another. Kirsch et al. (2012) note that this dynamic can lead to both positive and negative organizational experiences especially where organizations are seeking to enact change. Cook (1994), Shapira et al. (2010) and Manyard et al. (2012) all observe that there is a distinct leadership style which presents itself in Arab cultures and which is driven by long-standing societal beliefs and preferences that adduce patriarchal and authoritarian standards. In short, Arab culture is strongly hierarchical and collectivist in nature and this creates ramifications in organizational culture and leadership style (Skalli 2011).

The result is a patriarchal and hierarchical leadership style which is predominately authoritarian in nature. The evidence to date suggests that such a leadership style effectively blends into the Arab cultural backdrop in precisely the same manner that more empowered and participative leadership styles work effectively within a Western social construct. The challenge presents itself when the two directly conflicting styles are brought into proximity (Eastin & Prakash 2013). The recent Arab Spring has served to propel this situation into the popular domain and has highlighted the fundamentally different approaches to leadership and business which are present in Western and Arab cultures. At this juncture there is certainly no intention to suggest that one leadership style is superior to another, but rather to explore how and why they function in each context in order to elucidate a deeper understanding. Not until this understanding is gained will it be possible to assert whether changes in leadership style would bring value and benefit to organizations and their employees.

3.2.3 *Female leadership and empowerment*

Goby & Erogul (2011) continue the argument that female-run and -managed businesses have a tendency to form a matrix structure as opposed to a hierarchical pyramid. Since women are more communicative especially when under stress, as research in psychology has demonstrated (Lilienfeld et al. 2010), they are more effective at collating and then distributing useful information which can be shared amongst the organization members for collective benefit. It may be argued that since Eastern women do not seek to build power bases as a means of securing their positions in organizations, they do not accumulate information to use as an internal political lever. Arguably this is an alternative consideration of the suggestion that, because female leaders do not spend their time chasing power and prestige, it does not occur to them to withhold valuable information in the same manner as a male leader. The distribution of information and the fact that women are prepared to share their power more readily has the tendency to produce a more democratic and active organization which is better placed to respond to the demands of a dynamic marketplace. Moreover, because female leaders tend to have greater empathy with organizational stakeholders on an internal and external basis (Goby & Erogul 2011), they are better placed to respond rapidly to changes in the environment and to resolve issues before they become problems. The result is an empowered organization whereby employees have trust in their leader because they know that information is shared and they are working in an environment which focusses on growth and not accusation and reprimand for any errors that may occur.

3.3 *Quotas*

There is a wealth of research which examines the challenges female board members face, and what is labelled as the "casual" and "inadvertent" sexism they encounter. (One needs to clarify here that some feminist writers would consider the words "casual" and inadvertent" as offensive because for them no discrimination is casual; it comes out of deeply embedded and systemic bias.) Academics such as Bilimoria (2009), Ryan et al. (2011), Sealy & Vinnicombe (2010) and Seierstad & Healy (2012) have examined many dimensions of the roles of women in the boardroom and the challenges they encounter which range from holding lower-ranked roles (Bilimoria 2009) to having disproportionately fewer strategic positions and lesser corporate decision-making power (Sealy & Vinnicombe 2010). Even in regions where equality is actively promoted, such as Scandinavia, Seierstad and Healy (2012) have established that women in the workplace still face intangible challenges and subtle, casual discrimination leading to a disproportionate number of men holding the highest positions within organizations (Haslam & Ryan 2008, Ryan et al. 2011). Haslam & Ryan (2008) raised the question of whether women are sometimes set up to fail when they are recruited into positions of leadership in organizations that are known to be vulnerable to risks. In addition, even with the introduction of mandatory quotas for the employment of women in leadership positions, societal issues, deep-rooted beliefs, and misconceptions about women appear to stand in the way of practical application. In fact, those entrenched and systemic biases in society are sometimes so strong that, as was demonstrated in a study by Haslam & Ryan (2008), when companies fail, people think "female manager," and when companies succeed, people think "male manager."

The facts of the matter speak for themselves; there are more males than females in boardrooms and as of yet the imposition of positive quotas seems to have done little to redress the balance. There have been some suggestions that increasing positive discrimination within the workforce, and especially within the boardroom, would directly strengthen female empowerment (Greene & Kirton 2011). On the other hand, there are arguments against this notion since the idea is not merely to put token representations of females in the boardroom in order to raise the quota, but it is to promote deserving senior female employees (Horwitz et al. 2003). If one explores the issues more deeply, what seems to be needed is a paradigm shift in the thinking of society about females in business and in the boardroom. This needs further research and exploration and this study hopes to do that.

3.3.1 *The EU decision of November 2012*
Fontanella-Khan (2012) reported that the European Union (EU) made an announcement in November 2012 that as of 2020, the FTSE 100 and 250 organizations and their counterparts across Europe will be forced into recruiting female candidates into senior roles. This EU ruling goes further than anticipated, as it had been suspected that the ruling would simply enforce quotas with the aim of achieving 40% of female representation on the board. In addition, the EU has announced that organizations that do not recruit suitably qualified female candidates in preference to male candidates will face sanctions. It was noted by Evershed's, a leading global law firm, that "This [new rule] will require a comparative analysis of the qualifications of each candidate and the application of clear, gender-neutral and unambiguous criteria – an exercise which is fraught with risk and difficulties in practice."

Indeed it seems that this was partially the aim of the EU, which believes that the fear of potential litigation from female candidates will strongly encourage large companies to recruit females in preference to males. It is anticipated that in practice the legislation will require the organizations in question to demonstrate a clear policy of selection criteria, and if not selected, females will have the right to question and potentially challenge decisions. The EU believes that the increased tendency towards litigation in the employment arena for such scenarios will be sufficient to motivate large organizations to recruit women to senior positions in favour of men until they reach the 40% threshold.

Critics of the proposal argue that mandated quotas or administrative burdens in order to enforce female representation in the boardroom are a retrograde step. They point out that it is discriminatory against males and may well perpetuate the phenomenon of the glass cliff as previously discussed. They also argue that it will almost invariably lead to the appointment of women who do not necessarily have sufficient experience or are of the appropriate calibre to perform the roles. Some practitioners argue that female quotas or sanctions will be nothing but "tokenistic" (Peacock 2012a). When one examines such critics' arguments, however, one finds that they are feeble. This is because no such strict criteria have ever been applied to the male gender when discussing the latter's employment. In other words, has anyone ever stated that males should not be recruited in boardroom positions because, "perhaps," some will not have sufficient experience or some are "not of the appropriate calibre" to perform the job? This clearly demonstrates the existence of deeply rooted and systemic biases against women in societal organizations and structures.

Although recruitment of women who for whatever reason lack the necessary experience to carry out a senior board role is likely to create precisely the right conditions for them to fall from the glass cliff, according to Ryan et al. (2011), one needs to take into consideration that systemic discrimination has been going on for centuries and the only way to level the playing field is to make recruitment of females, whether in the boardroom or otherwise, mandatory. This explains why Blackburn & Jarman (2006) and also Peacock (2012a) determined in their research that even when support programmes such as fast tracking are in place, the necessary social support is often lacking; what can be concluded here is that the problem lies in the fact that society does not see or understand the need for that social support!

What is meant here is that mandatory quotas may help in getting women up the ladder, but this does not mean that all recruited females will excel; what it means, however, is that there will be enough females in the workplace to level the playing field, and through this enrichment and diversity of talent and contributions false generalizations and misconceptions about women are bound to be uprooted eventually and a more objective picture will emerge; women can weed out biases against them only when they are visible and through their pragmatic and practical work experience on a daily basis over time within organizations, and when they can articulate in their own voices their multiple concerns.

Initial evidence by professional organizations such as the Chartered Institute of Personnel and Development (CIPD) and the Chartered Management Institute (CMI) indicates that although when people enter the workforce (at approximately 18–25 years of age) they are almost equally split gender-wise, for a variety of reasons, as people progress through their careers, fewer than 50% of the females then progress beyond middle management. Some researchers maintain that there seems to exist no adequate explanation for this consistent trend of women failing to progress up the career ladder into senior executive positions (Lewis 2010). It is becoming clearer, however, that females are making conscious choices about climbing the corporate ladder; many seems to be laying aside opportunities to be in the boardroom in order to pursue entrepreneurial projects or in order to keep the balance of work and family intact, a balance they feel they will lose if they end up in leadership positions.

3.3.2 *Empowerment and work – life balance*

The term "work-life balance" was initially coined in the 1900s (Eastin & Prakash 2013). It refers to the situation whereby an individual is able to strike an appropriate balance between the amount of time spent in a low-paid working environment, and the amount of time that an individual spends on leisure or non-paid activities. There is a considerable body of evidence which demonstrates that employees who have a sensible work-life balance are in fact more productive because of the relaxation or downtime which they have built into their lives (Sturges & Guest 2004, Anderson & Eswaran 2009, Neumayer & De Soysa 2011; Eastin & Prakash 2013). To an extent there is an unwritten expectation that it is the responsibility of the employer to ensure that they do not overwork employees, and as noted by Anderson & Eswaran (2009), whilst this is relatively straightforward at the lower end of the employee spectrum, where employees work for fixed office hours or shifts, the more senior the employees the greater the expectation that they need to be available for communication out of office hours. This brings the discussion to the thorny matter of females balancing home and work commitments.

There is also evidence to suggest that it is at this point in their lives and careers that many women decide that they would rather put their skills and talents to entrepreneurial use and create their own businesses, which strikes an acceptable work-life balance for them (Allen et al. 2007, Coleman 2007, Anderson & Eswaran 2009). The A study by Blau & Kahn (2007) provides empirical support for the notion that once women have chosen to start a family, there is a sharp drop-off in the number of females in the economic sector. Large attrition rates referred to during the preceding discussions regarding female quotas are being picked up in much smaller entrepreneurial ventures and micro-businesses whereby women are able to strike a work-life balance which is acceptable to them.

Peacock (2012b) revealed that over 95% of women who would be capable of carrying out senior roles have actively chosen not to do so. Cumulatively the evidence would suggest that as women steadily progress up the career ladder they realize that the espoused and supposed benefits of senior executive roles are in fact not worth the sacrifices. In contradiction to pervading institutionalized biases against women, this finding demonstrates that women do think carefully about their decisions and make intelligent choices about their lives that do not necessarily prioritize their own career successes over the well-being of their families and spouses. Gilligan (1982) brought attention to the fact that when it comes to moral choices and choices that have to do with values, females think differently from males and have completely different priorities.

This evidence provides a link to the assertions of Bird & Brush (2002), Coleman (2007), Fairlie & Robb (2009), Gries & Naudé (2010) and Naudé & Rossouw (2010) who have all suggested that legislators have made a fundamentally flawed assumption in presuming that a large number of women actually want to progress to the boardroom. Recalling the opening discussions in this review and the definitions of empowerment which include the freedom to make choices, it is ultimately concluded that a significant proportion of females within the workplace may indeed be empowered in the sense of having agency and making their own decisions. This has manifested itself in a conscious choice not to progress up the career ladder in high-profile and large organizational roles. Instead, these women place greater personal and intrinsic value upon striking an acceptable work-life balance which satisfies their personal, intellectual, and family-based needs.

There needs to be a cautionary note here, however: one ought not to conclude that merely because women do not necessarily choose boardroom positions, that mandatory quotas should not be implemented. That would be a patronizing and false conclusion. In order to create a paradigm shift, level the field, and enable women who have different lifestyles, beliefs, ambitions, and personalities to have the widest spectrum of opportunities available, quotas would still need to be implemented mandatorily simply because that is the respectful procedure that will allow centuries of discrimination to be redressed. The bottom line is that no one should speak on behalf of women and no one ought to dictate to women what they should or should not do. Also, females need to be considered as equal to males as human beings, in terms of their capabilities, intelligence, talents, and their abilities to make the right moral choices.

The situation is very similar to discrimination against black people. When affirmative action was instituted, many people continued to speak on behalf of black people in a patronizing manner. Also, part of society started screaming "reverse discrimination" when mandatory quotas in the employment of minorities, especially blacks, were implemented in the United States. The truth and the reality are, however, that unless more minorities are effectively employed in the workforce, no-one will be able to tell the extent of their talents and achievements. When people are invisible, it is easy to usurp their rights.

3.3.3 *The female perspective*
Linking this back to the previous discussions of the actual numbers of women who had the requisite skills and expertise to take on senior executive roles, the fact that so many women drop out of the career path relatively early on provides a partial explanation for the very small available talent pool. Furthermore, a recent study reported by Peacock (2012b) has found that despite all of these efforts to induct women to the boardroom, over 95% of women surveyed, who would be capable of progressing to this level, declared that they actually did not want a board position. The study found that from the 547 females surveyed who would be capable of carrying out a board role, a mere 4% claimed that they had any aspirations for the board. The study found that 37% indicated that they wanted to reach a senior management level, but 17% indicated that they were perfectly happy to remain at a lower administrative level for the duration of their careers. Indeed, of all the females surveyed, 68% indicated that they would be perfectly happy with a mid-level role because it meant that they would be able to balance their work-life commitments. A further 27% stated that they felt that they would not be able to balance a senior executive role and a family and do justice to both, and 24% indicated that they felt that they would be ill-equipped to conduct work at board level. This confirms further the question of how women make their choices that Gilligan (1982) investigated decades ago. In other words, females do not prioritize their own successes over the welfare of others in their communities, especially if those others are family and children.

This evidence suggests that government quotas and EU directives are failing to address some fundamental questions which have been raised by academics over the last decade. More worryingly it would seem that some very significant assumptions about women and about their capabilities have been drawn on unsubstantiated claims. In other words, what is transpiring is that it is not because females are incapable, not talented, or not endowed with leadership qualities that they are not climbing the corporate ladder, it is because they are making certain conscious choices. Whilst the EU presses ahead with female quotas or sanctions for organizations that fail to meet the expectations of placing women on boards, it would seem that more complex issues are at play in a significant proportion of cases; perfectly capable women have in fact chosen not to join the senior executive of a large company. Arguably, the statistics presented here support the research work of scholars such as Brush (1992), Bird & Brush (2002), Fairlie & Robb (2009), Gries & Naudé (2010), Naudé & Rossouw (2010) and Coleman (2007) all of whom have comprehensively demonstrated over a 20-year period that females are most capable of running organizations. What is telling is that all of these organizations that are run by women have deliberately chosen to remain in the small-to-medium category. The body of research in this area indicates that small to medium enterprises (SMEs) typically engender a different type of organizational culture, which is often much less bureaucratic than that which is associated with large multinational corporations (MNCs) (Barney 1986, Hartnell et al. 2011). Linking this finding back to the previous discussions of the so-called "glass cliff" (Hladik 2012), it would suggest that women are well aware of the hurdles that they would have to overcome as members of the boardroom within patriarchal and hierarchical organizations and they choose, instead, to build organizations which remain small and cohesive with a highly collaborative culture.

3.3.4 *The evidence and arguments for and against quotas*
There is continued heated debate as to whether or not female quotas should be enforced in senior executive boardrooms; the debate is being grounded in the fact that there is some scope for negotiation in the wording of the EU ruling. Although some regions such as Scandinavia have had considerable success with quotas, other countries have deliberately not sought to enforce them. Again, it is worth reiterating that the positive experiences of Scandinavia are largely attributed to underlying cultural and social norms which are much more progressive and understanding of women's thinking and choices. In the light of the preceding discussions there needs to be concerted effort to explore the root causes of female disempowerment which have to do more with prevalent misconceptions about female capabilities, ubiquitous biases about women's needs and wants, and a lack of seriousness in exploring scientific psychological research about women. These may be key factors in articulating a more objective discussion about the dearth of females in boardroom positions which will eventually, through proper scaffolding of critical issues, shed more light on whether the enforcement of quotas would present a positive benefit over time.

In this literature review, what the authors are articulating is that no single law or regulation can redress the imbalance and restore to women their right of full empowerment; moreover, what the authors are attempting to highlight is that a more holistic and comprehensive discussion is needed that examines women's issues and life situations from multiple perspectives, and incorporates much more research by women about women and expressed authentically through women's voices. Furthermore, it would be naive to presume that the introduction of female quotas would produce an immediate benefit. Indeed, it might take at least a decade, if not a generation, for the benefits to filter through, because quotas must be supported by a cultural and societal paradigm shift, which the authors are suggesting considering the above discussion.

Changing hierarchical and patriarchal thinking is not easy; there are in society as many misconceptions about the realities of males as there are about females. Without this intellectual shift in thinking about women, it is very hard to look at issues relating to women from a non-biased perspective since that bias has been entrenched globally across ages. Hence, countries that refrain from enforcing quotas need to clarify why they are doing that. The argument that they are not enforcing quotas in order to protect women from scaling the glass cliff is patronizing. Women are perfectly capable of protecting themselves and of speaking for themselves.

There is also empirical evidence to suggest that organisations with mixed gender boards have better financial success in the long-term, which is attributed to factors such as females being risk-averse in a long-term context, meaning that they take a longer view and do not attempt to grow the company too quickly in pursuit of quick profit and high risk.

It stands to reason to question whether the excuse of not introducing quotas is legitimate when the reason is fear of recruiting the wrong females who are perhaps not qualified!! Is that a logical argument? Indeed, previously in this area of research, it has been observed that females who were already in leadership positions who achieve these roles as a result of their own merit and hard work are often considered to be "the token female" even when quotas have not been introduced. It is suggested that women who seek leadership roles, certainly wish to achieve leadership on the basis of merit and not patronage (Ryan & Haslam 2007a, Sealy & Vinnicombe 2010). Some researchers fear that women who have worked very hard in their lives to achieve leadership roles are likely to be challenged by the introduction of quotas since it might give the impression that their positions were achieved through quotas rather than through their own work; this is an unreasonable fear since they can always demonstrate their capabilities through experience on site! Finally, it could happen that in the introduction of mandatory quotas, some qualified males might be disregarded at the expense of hiring women in order to adhere to the minimum quota requirements. Sometimes some compromises need to be made in order to rectify a situation that has been flawed for a long time.

Even though research in many Scandinavian countries explores the limitations of the imposition of quotas (Pande & Ford 2011), one would be naïve not to suggest that, first, more females need to be recruited into the workplace and in leadership positions in general; second, much more research is needed to unpack the issues and explore the advantages and disadvantages of quotas (Pande & Ford 2011, Elstad & Ladegard 2012); third, through education about women, their history, their struggles, and how they came to be so disadvantaged, different sectors of society will cease to speak on behalf of women and will attempt to give equal opportunities for them, especially in the domain of training and the acquisition of skills for leadership positions (Villiers 2010). Finally, more research is needed to explore any other factors that may underpin the perceived lack of success of women in boardroom positions (Ryan & Haslam 2007a,b; Lazzarettiet al. 2013).

Attempting to force societal change through legislation is, at best, a short-term measure; and women definitely do not want to be perceived as the recipients of charity or "tokenism." The consequences of not imposing quotas in the long term and in the macro picture are, however, much more damaging than the consequences of imposing quotas; the authors maintain that the talents and capabilities of women cannot continue to be hidden; they need to become transparent, especially at the boardroom level. How can there be more exploration of how females handle top executive positions if organizations and governments do not open the doors wide for them? In order to level the playing field, sometimes positive discrimination is needed temporarily. More importantly, inter-and multi-disciplinary research is needed to effectively gauge the authentic reasons that drive women to make certain choices that are different from those men make regarding boardroom positions and leadership roles in general. Reality is socially constructed, as Berger & Luckman (1966) famously asserted, and the truth is multi-dimensional.

4 PROPOSED FUTURE RESEARCH

Many researchers have addressed the concepts of empowerment of women, leadership aspects related to women and quotas using various research methodologies. For instance, Goodman et al. (2003) used quantitative research methodology to investigate the topic related to female representation in organizations. In contrast Abu-Rabia-Queder & Oplatka (2008) have used qualitative research methodology to gain in-depth knowledge regarding the use of quotas in the corporate boardroom and whether quotas favour women or otherwise. An examination of the research conducted by those authors indicates that the research methodology used by them is dictated by the research questions addressed by them. In this research the purpose is to gain an in-depth understanding of how quotas could be used to favour women and empower them to occupy boardroom positions and contribute with their leadership abilities within organizations.

In addition, understanding the phenomenon of quotas as a related literature has not produced theories or models or conceptualization that could be applied to understand the phenomenon. Therefore, it is essential to study the perceptions of individuals working in the corporate sector. Such a study needs to involve qualitative research methodology, for instance, conducting semi-structured interviews to collect qualitative data, an argument supported by other researchers (Kumra & Vinnicombe 2010).

More importantly, linking the concept of quotas to gender barriers to find resolutions to overcome the barriers has not been investigated well in the literature. An investigation into this aspect requires understanding the behaviour of people, their perceptions, experiences, feelings, thoughts and ideas, so that a framework could be developed to apply quotas in organizations for overcoming gender discrimination. Moreover, this study involves a data sample of 100 participants. The sampling would be done based on a random sampling method to get feedback via survey/questionnaires from some of society's management sectors about the mandatory implementation of quotas in the boardroom. In statistics the random sampling corresponds to a draw from a population distributed according to the law (density function) of a given number of individuals/objects. The sample selection in the random sample is assigned to the case and should not be influenced, consciously or unconsciously, by those who carry out the investigation. The essential features of simple random sampling are

a) all units in the population have an equal chance to be part of the sample,
b) each sample of size n has the same probability to be formed.

The key participants from whom data would be collected are females from Bahrain's private sector who occupy significant positions within organizational structures including the corporate boardroom. This part of the research is in progress.

5 CONCLUSIONS

This paper has addressed three discrete but linked areas of research in regard to the nature of female empowerment, female leadership and quotas for women on various organizational levels with a focus on the corporate boardroom. The forgoing discussions clearly demonstrate that the current level of research and knowledge in the areas of empowerment, leadership, and quotas for women at the boardroom level are inadequate to create a comprehensive and cohesive critical discussion about those topics. Although a literature review shows that some researchers have highlighted the need to implement quotas within organizations at multiple levels (Rai 2012), little is known about the effectiveness of enforcing quotas for the female labour force at the boardroom level and whether quotas would, in the long run, empower them or otherwise (Pande & Ford 2011). That is not least because it was discovered that many women decline opportunities to climb the glass cliff even if they are able to because their priorities are different from men and because they need to create a work-life balance. It is apparent from the foregoing literature review that the investigation of this question needs to be done within the larger context of societal entrenched systemic biases against women and the lack of enough education about how women think and make decisions. It also needs to be done in the context of women speaking for themselves, particularly in the Middle East and in places where there are no quotas introduced at all. Thus, instead of asking the question of whether the introduction of mandatory quotas at the boardroom level will empower women and make them effective leaders, one needs to ask the following questions:

1. Do women face the same barriers as males in the workplace as they attempt to achieve top management positions?
2. What are some of the barriers that exist within the workplace and in society at large that prevent women from achieving boardroom positions?
3. What type of coping mechanisms such as quotas/empowerment could be implemented to reduce the barriers that prevent women from achieving positions in the boardroom?

In order to answer these questions, the authors suggest that semi-structured interviews be conducted with women who are already in the workplace, whether in management or non-management positions, to gauge how they evaluate their own empowerment or lack thereof.

REFERENCES

Abbott, N.G., White, A.F. & Charles A.M. 2005. Linking values and organizational commitment: A correlational and experimental investigation in two organizations. *Journal of Occupational and Organizational Psychology* 78(32): 301–321.

Abu-Rabia-Queder, S. & Oplatka, I. 2008. The power of femininity: Exploring the gender and ethnic experiences of Muslim women who accessed supervisory roles in a Bedouin society. *Journal of Educational Administration* 16(3): 396–415.

Allen, E., Langowitz, N. & Minniti, M. 2007. *The 2006 Global Entrepreneurship Monitor Special Topic Report: Women in Entrepreneurship*. NY: Center for Women's Leadership Babson Park, Babson College.

Anderson, S. & Eswaran, M. 2009. What determines female autonomy?: Evidence from Bangladesh. *Journal of Development Economics* 90: 179–191.

Barney, J.B. 1986. Organizational culture: Can it be a source of sustained competitive advantage? *Academy of Management Review* 11(3): 656–665.

Bilimoria, D. 2009. *Women on Corporate Boards of Directors: International Research and Practice*. Cheltenham: Edward Elgar Publishing.

Bird, B.J. & Brush, C.G. 2002. A gendered perspective on organizational creation. *Entrepreneurship Theory and Practice* 26(3): 41–65.

Blackburn, R.M. & Jarman, J. 2006. Gendered occupations: Exploring the relationship between gender segregation and inequality. *International Sociology* 21(2): 289–315.

Blau, F.D. & Kahn, L.M. 2007. Changes in the labor supply behavior of married women: 1980–2000. *Journal of Labor Economics* 25(3): 393.

Brescoll, V.L., Dawson, E. & Uhlmann, E.L. 2010. Hard won and easily lost: The fragile status of leaders in gender-stereotype-incongruent occupations. *Psychological Science* 21(1): 1640–1642.

Bruckmüller, S. & Branscombe, N.R. 2010. The glass cliff: When and why women are selected as leaders in crisis contexts. *British Journal of Social Psychology* 49(3): 433–451.

Brush, C.G. 1992. Research on women business owners: Past trends, a new perspective and future directions. *Entrepreneurship Theory and Practice* 16(4): 5–31.

Burn, S.M. 2005. *Women Across Cultures: A Global Perspective*. New York, NY: McGraw-Hill.

Coleman, S. 2007. The role of human and financial capital in the profitability and growth of women-owned small firms. *Journal of Small Business Management* 45(3): 303–319.

Collins, D. 1994. The disempowering logic of empowerment. *Empowerment in Organizations* 2(2): 14–21.

Cook, S. 1994. The cultural implications of empowerment. *Empowerment in Organizations* 2(2): 9–13.

Cooney, R. 2004. Empowered self-management and the design of work teams. *Personnel Review* 33(1): 677–692.

Cotter, D.A., Hermsen, J.M., Ovadia, S. & Vanneman, R. 2001. The glass ceiling effect. *Social Forces* 80(2): 655–681.

Cox, A., Zagelmeyer, S. & Marchington, M. 2006. Embedding employee involvement and participation at work. *Human Resources Management Journal* 16(3): 250–267.

Datta, R. & Kornberg, J. (eds) 2002. Introduction in *Women in Developing Countries: Assessing Strategies of Empowerment*. London: Lynne Rienner.

Davies-Netzley, S.A. 1998. Women above the glass ceiling: Perceptions on corporate mobility and strategies for success. *Gender and Society* 12(3): 340–345.

Denham-Lincoln, N., Travers, C., Alimo-Metcalfe, 1995s, P., Wilkinson, A. 2002. The meaning of empowerment: The interdisciplinary etymology of a new management concept. *International Journal of Management Reviews* 4(3): 271–290.

Eagly, H.A., Johannesen, C.M. & Engen, L.M. 2003. Transformational, transactional and laissez-faire leadership tyles: A meta-analysis comparing women and men. *Psychological Bulletin* 129(4): 569–591.

Eastin, J. & Prakash, A. 2013. Economic development and gender equality: Is there a Gender Kuznets Curve? *World Politics* 65: 1.

Elstad, B. & Ladegard, G. 2012. Women on corporate boards: Key influencers or tokens? *Journal of Management and Governance* (in press).

Fairlie, R.W. & Robb, A.M. 2009. Gender differences in business performance: Evidence from the characteristics of business owners' survey. *Small Business Economics* 33(4); 375.

Federal Glass Ceiling Commission (FGCC) 1995. *Solid Investments: Making Full Use of the Nation's cman Capital*. Washington, DC: U.S. Department of Labor, November 1995, 4.

Fleming, P. & Spicer, A. 2003. Working at a cynical distance: Implications for power, subjectivity and resistance. *Organization* 10(1): 57–180.

Fleming, P. & Spicer, A. 2007. *Contesting the Corporation*. Cambridge, UK: Cambridge University Press.

Fontanella-Khan, J. 2012. Sanctions to enforce female board quotas [on-line 14 November 2012]. *Financial Times* [retrieved 28 November 2012]. http://www.ft.com/cms/s/0/4621efb6-2e80-11e2-9b98-00144feabdc0.html#axzz2DVGnsrBG

Gajjala, R., Zhang, Y. & Dako-Gyeke, P. 2010. Lexicons of women's empowerment online. *Feminist Media Studies*, 10(1): 69–86.

Gilligan, C. 1982. *In a Different Voice*: *Psychological Theory and Women's Moral Development*. Cambridge, MA: Harvard University Press.

Goby, V.P. & Erogul, M.S. 2011. Female entrepreneurship in the United Arab Emirates: Legislative encouragements and cultural constraints. *Women's Studies International Forum* 34(4): 329–334.

Goodman, J.S., Fields, D.L. & Blum, T.C. 2003. Cracks in the glass ceiling: In what kind of organisations do women make it to the top? *Group Organisation Management* 28(4): 475–501.

Greene, A. & Kirton, G. 2011. The value of investigating stakeholder involvement. In G.S. Farnham, *Diversity Management. Diversity in the Workplace: Multi-disciplinary and International Perspectives*. Bristol: Gower.

Grey, C. 1999. We are all managers now? We always were: On the development and demise of management. *Journal of Management Studies* 36(5): 561–586.

Gries, T. & Naudé, W.A. 2010. Entrepreneurship and structural economic transformation. *Small Business Economics Journal* 34(1): 13–29.

Guillaume, Y.R.F. & Telle, N.-T. 2011. Leading and influencing in organisations. In M. Butler & E. Rose (eds), *Introduction to organisational behaviour*. London: CIPD, pp. 205–236.

Hales, C. 2001. Does it matter what managers do? *Business Strategy Review* 12(2): 50–58.

Hartnell, C.A., Ou, A.Y. & Kinicki, A. 2011. Organizational culture and organizational effectiveness: A meta-analytic investigation of the competing values framework's theoretical suppositions. *Journal of Applied Psychology* 96(4): 694–706.

Haslam, S.A., and Ryan, M.K., (2008) The road to the glass cliff: Differences in the perceived suitability of men and women for leadership positions in succeeding and failing organizations. *Leadership Quarterly* 19(1) 530–546.

High-Pippert, A. & Comer, J. 1998. Female empowerment. *Women & Politics* 19(4): 53–66.

Hinkin, T.R. 1995. A review of scale development practices in the study of organizations. *Journal of Management* 21(5): 967–988.

Hinkin, T.R. & Schriesheim, C.A. 1989. Development and application of new scales to measure the French and Raven (1959) bases of social power. *Journal of Applied Psychology,* 74(4): 561–567.

Hladik, M. 2012. Grass ceiling: Overhangs surge in female farmers [online]. *Women's News* available at http://womensenews.org/story/entrepreneurship/120919/grass-ceiling-overhangs-surge-in-female-farmers [retrieved 27 November 2012].

Horwitz, F.M., Chan, T.-H. & Quazi, H.A. 2003. Finders, keepers? Attracting, motivating and retaining knowledge workers. *Human Resource Management Journal* 13(4): 23–44.

Hu, J., Wang, Z., Liden, R.C. & Sun, J. 2012. The influence of leader core self evaluation on follower reports of transformational leadership. *The Leadership Quarterly* 10(1): 1048.

Jonsen, K., Martha L., Maznevski, S. & Schneider, C. 2010. Gender differences in leadership – believing is seeing: Implications for managing diversity. *Equality, Diversity and Inclusion* 29(6): 549–572.

Kirsch, C., Chelliah, J. & Parry, W. 2012. The impact of cross-cultural dynamics on change management. *Cross Cultural Management* 19(2): 166–195.

Knudsen, K. & Wærness, K. 2009. Housework in thirty-four countries. *The International Social Survey Programme*, 1984–2009: *Charting the Globe* 378.

Kumra, S. & Vinnicombe, S. 2010. Impressing for success: A gendered analysis of a key social capital accumulation strategy. *Gender Work and Organization* 17(5): 521–546.

Lee, M. & Koh, J. 2001. Is empowerment really a new concept? *International Journal of Human Resource Management* 12(4): 684–695.

Lewis, S. 2010. Restructuring workplace cultures: The ultimate work-family challenge? *Gender in Management* 25(5): 355–365.

Lazzaretti, K., Godoi, C.K., Camilo S.P.R. & Marcon, R. 2013. Gender diversity in the boards of directors of Brazilian businesses. *Gender in Management* 28(2): 94–110.

McDowell, J.M., Singell, L.D. & Ziliak, J.P. 1999. Cracks in the glass ceiling: Gender and promotion in the economics profession. *American Economic Review* 82(2): 392–396.

Maynard, T., Gilson, L.L. & Mathieu, M.E. 2012. Empowerment – Fad or fab? A multi-level review of the past two decades. *Research Journal of Management* 38(4): 1231–1281.

Moghadam, V.M. 2003. *Modernizing Women: Gender and Social Change in the Middle East*. Boulder, Colorado & London: Lynne Reiner Publishers.

Naudé, W.A. & Rossouw, S. 2010. Early international entrepreneurship in China: Extent and determinants. *Journal of International Entrepreneurship* 8(1): 87–111.

Neumayer, E. & De Soysa, I. 2011. Globalization and the empowerment of women: An analysis of spatial dependence *via* trade and foreign direct investment. *World development* 39(7): 1065–1075.

Pande, R. & Ford, D. 2011. *Gender Quotas and Female Leadership*: *A Review*. Background paper for the World Development Report on Gender.

Pastor, J. 1996. Empowerment: What it is and what it is not. *Empowerment in Organizations* 4(2): 5–7.

Peacock, L. 2012a. We need female quotas because men aren't listening. *The Telegraph* [on-line, 23 October 2012]. Aavailable at http://www.telegraph.co.uk/women/womens-business/9628429/We-need-female-board-quotas-because-men-arent-listening.html [retrieved 28 November 2012].

Peacock, L. 2012b. Listen up EU "many women don't want to work on boards". *The Telegraph* [on-line 23 October 2012]. Available at http://www.telegraph.co.uk/finance/jobs/9626408/Listen-up-EU-many-women-dont-want-to-work-on-boards.html [retrieved 28 November 2012].

Peeters, G. 1991. Evaluative inference in social cognition: The roles of direct *versus* indirect evaluation and positive-negative asymmetry. *European Journal of Social Psychology* 21(2): 131–146.

Purcell, J. & Hutchinson, S. 2003. *Bringing Policies to Life*: *The Vital Role of Front Line Managers in People Management*. London: CIPD.

Quinn, R. & Spreitzer, G. 1997. The road to empowerment: Seven questions every leader should consider. *Organizational Dynamics* 26(2): 37–49. [Business Source Premier, EBSCO*host*, viewed 22 March 2013.]

Rai, S. 2012. Female presence in boardrooms: Review of global scenario. *Asian Journal of Management Research* 3(1).

Rodrigues, C.A. 1994. Employee participation and empowerment programs: Problems of definition and implementation. *Empowerment in Organizations* 2(2): 29–40.

Ruth, S. 1990. *Issues in Feminism: An Introduction to Women's Studies*. London & Toronto: Mayfield Publishing Company.

Ryan, M.K. & Haslam, S.A. 2004. Beyond the glass ceiling. Women in the boardroom: A bird's eye view. *CIPD Change Agenda*. London: CIPD.

Ryan, M.K. & Haslam, S.A. 2005a. The glass cliff: Evidence that women are over-represented in precarious leadership positions. *British Journal of Management* 16(1): 81–90.

Ryan, M.K. & Haslam, S.A. 2005b. The glass cliff: Implicit theories of leadership and gender and the precariousness of women's leadership positions. In B. Schyns & J.R. Meindl (eds), *Implicit Leadership Theories: Essays and Wxplorations*. Greenwich, CT: Information Age Publishing, pp. 137–160.

Ryan, M.K. & Haslam, S.A. 2007a. The glass cliff: Exploring the dynamics surrounding the appointment of women in precarious leadership positions. *Academy of Management Review* 32(1): 549–572.

Ryan, M.K. & Haslam, S.A. 2007b. The glass cliff: The risks of being on top. *CIPD Change Agenda*. London: CIPD.

Ryan, M.K., Haslam, S.A., Hersby, M.D. & Bongiorno, R. 2011. Think crisis – think female: The glass cliff and contextual variation in the think manager – think male stereotype. *Journal of Applied Psychology* 96(1): 470–484.

Sealy, R. & Vinnicombe, S. 2010. Boardroom balance. *Management Focus* (28): 24–25.

Seierstad, C. & Healy, G. 2012. Women's equality in the Scandinavian academy: A distant dream. *Work, Employment and Society* 26(2): 296–313.

Shapira, T., Khalid, A. & Azaiza, F. 2010. Arab women principals' empowerment and leadership in Israel. *Journal of Educational Administration* 48(6): 704–715.

Singh, V. & Vinnicombe, S. 2004. Why so few women directors in top UK boardrooms? Evidence and theoretical explanations. *Corporate Governance* 12(4): 479–488.

Skalli, L. 2011. Constructing Arab female leadership lessons from the Moroccan media. *Gender & Society* 25(4): 473–495.

Smircich L. 1983. Concepts of culture and organizational analysis. *Administrative Science Quarterly* 28(1): 339–358.

Stanley, G. 1994. Organisational culture and individual sensemaking: A schema-based perspective. *Organisation Science* 5(3): 309–321.

Sturges, J. & Guest, D. 2004. Working to live or living to work? Work/life balance early in the career. *Human Resource Management Journal* 14(4): 5–20.

Villiers, C. 2010. Achieving gender balance in the boardroom: Is it time for legislative action in the UK?. *Legal Studies* 30: 533–557.

Whiteside, M., Tsey, K. & Earles, W. 2011. Locating empowerment in the context of indigenous Australia. *Australian Social Work* 64(1): 113–129.

Willmott, H. 1984. Images and ideals of managerial work: A critical examination of conceptual and empirical accounts. *Journal of Management Studies* 21(3): 349–368.

Yammarino, F.J., Spangler, W.D., High-Pippert, A. & Comer, B.M. 1993. Transformational leadership and performance: A longitudinal investigation. *Leadership Quarterly* 4(1): 81–102.

Evaluating professional development in healthcare with outcome based models

Litty M. Shibu & Ebrahim K. Rajab
Ahlia University, Kingdom of Bahrain

Tillal Eldabi
Brunel University, UK

ABSTRACT: The objective of this paper is to review the concept of continuing professional development (CPD) in physical therapy. Outcomes of CPD, evaluation models of CPD, and effectiveness of CPD in changing the practice behaviour of the physical therapist to enhance evidence based practice (EBP) are discussed. The evolution of continued medical education (CME) to CPD as well the importance of CPD in the field of physical therapy is outlined. The available outcome based evaluation models are compared and the need for such a model for evaluating effectiveness of CPD is recognized. A research gap is identified as to how CPD can enhance EBP by changing the practice behaviour of physical therapists.

1 INTRODUCTION

Continuing Professional Development (CPD) is often referred as an important component to maintain professional competency across various professions (Alsop 2000). Entry level education does not prepare the professional with all the necessary knowledge, skills or attitude (Eraut 1994) to deal with the complex situations while advancing through one's career until retirement. Many reasons, like rapid expansion of available knowledge base, fast developments in technology and new approaches of client centred service delivery make it necessary for the professional to engage in learning throughout professional life (Ahuja 2011, French & Dowd 2008). The concept of CPD essentially translates the idea that all professionals need to engage in a continuous learning to keep themselves up-to-date with knowledge, techniques, revisions, modifications and developments, in tandem with the dynamics of every profession. Particularly in healthcare, it is the need of the hour for providing the best patient care based on the latest evidence available. In the field of physiotherapy research findings indicate that there is an augmented demand to continually update skills, knowledge and latest techniques for quality patient care. Moreover, the statutory requirement for revalidation of licence mandates participation in CPD in many countries (PBA 2013, CSP 2006). Even though there are great variations regarding the elucidation of CPD, there is still consensus in the literature regarding the common characteristics of CPD as a technique for updating knowledge, enhancing skills, maintaining professional competency for the overall development of the professional in a dynamic healthcare environment (PBA 2013).

Participation in CPD, as a tool for continuing competence of healthcare professionals, needs further investigation. Furthermore it is not yet clear about the role of CPD, to enhance competence and to promote the evidence based practice. Professionals as well as organizations as sponsors of CPD need further clarity regarding the cost benefit of CPD that could be related to specific outcomes, for promoting sponsorships for CPD activities. Interestingly the arguments of some researchers highlight the lack of in depth research of CPD as a concept in general and specific to healthcare profession. With reference to the field of physiotherapy there is general apathy towards CPD (Landers et al. 2005), that acts as barrier for physiotherapists to engage in CPD activities; to reap the benefits that it offers. Physical therapy (PT) as a profession has been privileged with autonomy of practice for more than five decades. The current scenario is, however, such that most of the conclusions regarding CPD are drawn from medical, nursing or other professions (French & Dowd 2008, Ahuja 2011, Chipchase et al. 2012). Hence there is a need for further research into the role of CPD, so as to enhance physical therapists' development as well as the overall development of the profession. Therefore, this review intends to outline the effectiveness of CPD in healthcare using outcome based evaluation.

1.1 Search strategy for literature review

The aim of the search strategy was to identify the effectiveness of CPD using evaluations of the outcomes, using multiple key words: continuing medical education, continuing professional development, lifelong learning, effectiveness, evaluation, impact, outcomes, physiotherapy, physical therapy, collaborative learning, and evidence based practice. An electronic search was conducted using the following databases: Medline, CINHAL, PEDro (Physiotherapy Evidence Database), using the key words identified for the purpose of this review. The search was restricted to articles published only in English (see Figure 1).

Each database was searched using the Boolean connectors, 'AND' and 'OR', to create a thorough search for better results. For example, continuing professional development 'AND' effectiveness 'AND' physiotherapy: (1) inclusion criteria as the published empirical research specific to the physiotherapy profession for the period of 1980–2014 and (2) exclusion criteria as content not directly related to continuing professional development related to the physiotherapy profession.

2 LITERATURE REVIEW

2.1 Continuing medical education

It is not possible for professionals to practice the same things in the same way for the whole of their working life ignoring the evolving nature of professional practices. Dubin (1972) proposed the half-life concept of medical knowledge, which states that the acquired medical knowledge becomes obsolete in about five years' time, which implies that the knowledge and professional competence are perishable commodities lasting between 2 to 5 years and there is a need to advance one's competence through continuous development. The first recorded continuing education in the field of nursing dates back to the eighteenth century (Stein 1998). Medical professionals participated in traditional didactic instructional methods such as lectures, conferences and seminars activities for decades. Later on continuing education for medical professionals were made mandatory in the USA during the 1970s and still exists as mandatory requirement for revalidation of the practice licence.

While healthcare professionals engage in continuing education, there was general apathy towards this in the field of physiotherapy until recently. During the 1980s, physiotherapy as a profession started recognizing continuing education and it was made mandatory in selected states in the USA (French & Dowd 2008). Although the profession outlined a rationale for professionals to engage in continuing education activities, evidence suggests that engagement in continuing education is not common in physical therapy and often participation in continuing education is largely for attaining required credits for the revalidation of a licence (Ahuja 2011, Chipchase et al. 2012, French & Dowd 2008). This indicates the attitude and behaviour of healthcare professionals towards continuing education. The literature highlights the need for a comprehensive and unified system of continuing education in the health professions beyond the knowledge gain, but as a method for addressing performance inadequacies of the professional as well as the overall healthcare systems level (Institute of Medicine 2001, 2009).

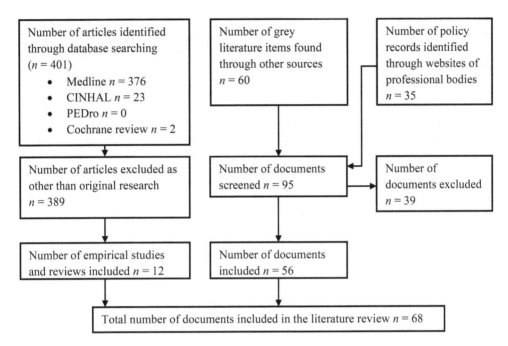

Figure 1. Flow chart for selection of articles for the literature review.

2.2 *Moving beyond continuing medical education*

The evolution of the idea that the knowledge which is acquired through entry level courses is not sufficient to ensure continued competence, ultimately resulted in the adoption of continuing education (Clyne 1995). The last few decades witnessed the emergence of a broader concept called 'Continuing professional development (CPD)' that often replaces the term continuing education. Although these terms are used interchangeably in the literature, there are obvious differences existing among them. CPD is viewed as an ongoing process and not as an end product (PSI 2010). CPD encompasses wider dimensions of the comprehensive development of the professional, whereas continuing education can be viewed as one of the methods to achieve that particular goal. Although it is an emerging concept, many professional bodies in the developed countries promote participation in CPD. Moreover, many of the developed countries like the UK, the USA, New Zealand and Australia have already made this mandatory as CPD of specified hours or credits are required for compliance professional practice licensing in healthcare (French & Dowd 2008). Laws and jurisdiction regulate participation in CPD, in the context of developed countries, whereas the driving factors that promote CPD in the developing countries in the absence of a mandatory requirement is yet to be explored.

2.3 *Concept of CPD*

Definitions of CPD vary across the professions. For instance in the field of education CPD has been described as "the maintenance and enhancement of the knowledge, expertise and competence of professionals throughout their careers according to a plan formulated with regard to the need of the professional, the employer, the profession and society" while in the context of healthcare, the Chartered Society of Physiotherapy, UK, defines CPD as "the process through which individuals undertake learning, through a broad range of activities that maintains, develops, and enhances skills and knowledge in order to improve performance in practice" (CSP 2007). It is evident from the above-mentioned definitions that even though CPD includes components of continuing education it has a broader potential. CPD is expected to integrate the multi-dimensional aspects of development of an individual far beyond mere knowledge and the skills required for professional practice. CPD includes personal and professional development which, in fact, needs to be mirrored on actual clinical practice (Nolan et al. 1995).

2.4 Characteristics of CPD

The literature suggests that CPD is regarded as a learner-driven activity, where the need identification done by the professional, who would be most likely to be aware of the own need for further learning. Hence, CPD essentially highlights that the control of learning and development is conferred upon all health professionals. Subtly it provides the flexibility to adjust to the learning needs of individual professionals, thereby empowering them to be in control of their own learning pace (WFME 2003). CPD is regarded as a continuous process, as Eraut (1994) points out that qualification should not be an end point of the learning process, but rather the beginning of professional development. Longworth & Davies (1996) support this view as lifelong learning has a personal and an educational meaning that the learning decisions are made by the professional, throughout the career, in order to ensure overall personal and professional development, that spans from cradle to grave.

CPD as a concept is founded on a broader set of theories like adult learning theory, reflective practice and situated learning (Institute of Medicine 2009). Moreover CPD identifies an extensive variety of learning methods such as self-directed learning, reflective learning, and collaborative learning. CPD also provides learning opportunities spanning from structured and planned formal activities to informal experiential learning that occurs incidentally and unplanned in the workplace (Dowds & French 2008). According to many researchers 70% of the learning takes place in an informal work-based environment (Enos et al. 2003). Most of the studies are, however, limited to formal structured CPD (Petty & Morley 2009) while traditional instructional methods are already being questioned in the literature for their ability to improve the participant's knowledge, skills, etc. (Mansouri & Lockyer 2007, Ahuja 2011). It is clear that there is a paucity of literature regarding informal CPD that may be due to lack of awareness of professionals to recognize the opportunity being offered by informal activities.

2.5 CPD in healthcare

Acknowledgment of CPD in healthcare is evident when it was made a statutory requirement to ensure continued competence. For instance, the proposed frameworks of various professional bodies like the British Medical Association and the American Medical Association, have adopted the concept as a guide for multi-faceted, lifelong learning and development of the healthcare professional. CPD is expected to be more beneficial than traditional continuing education if the activities are planned, implemented and evaluated in a systematic manner (Institute of Medicine 2009). As far as the benefits of CPD are concerned, it is meant to improve the participant's knowledge, skills and attitude (Davis et al. 1999, 2003). CPD should result in improved competence of the healthcare professionals who underwent the training, to provide the best possible patient care and improve patient outcomes, as well as the safety of patients (Umble & Cervero 1996). The question remains, however, 'Can CPD contribute to continued competence in the healthcare profession?' This is debatable. Even though it is believed that CPD should enhance the competence of the professional and improve the quality of care, CPD providers and participants focus on the purpose of attaining necessary credits, as it is largely viewed as an industry focussing on formal course-based activities (O'Brien et al. 2001).

2.6 CPD in physical therapy

Physical therapists are allied healthcare professionals who have entrusted with the responsibility of patient care with autonomy of practice (FSPT) since 1978 in the USA. An implication of this autonomy is that physical therapists are expected to continually update themselves in accordance with the dynamics of the profession by engaging in CPD activities (CSP 2013). Qualification as a physiotherapist should not be considered as the end point of an educational process, rather the beginning of an ongoing professional development through self-directed learning (Gosling 1999). Internationally, many of the statutory and professional bodies of PT (physical therapy) mandate their professionals to engage in the minimum requirement of CPD, to have compliance for registration and revalidation purposes. For instance, professional organizations like the American Physical Therapy Association (APTA), the Chartered Society of Physiotherapy (CSP), the Physiotherapy Board of Australia (PBA), promote CPD to enhance their skills and knowledge to contribute to overall effectiveness of patient care (French & Dowd 2008). Even though CPD as a concept was accepted in physical therapy, the approach to CPD planning, implementation and evaluation has a fragmented approach that has resulted in a lack of awareness and enthusiasm among professionals (Chipchase et al. 2012). The World Confederation of Physical Therapy (WCPT 2011), however, has recognized CPD and provided necessary guidelines urging member organizations to develop frameworks in accordance with their regional needs for maintaining and/or increasing the physical therapist's competence through CPD.

2.7 Research concerning CPD in the physical therapy profession

Continuing professional development as a topic has attracted the imagination of many researchers in the healthcare sector and more recently in the field of physical therapy. Table 1 presents the summary of empirical research conducted on CPD in the field of Physical therapy. Although many researches are conducted on CPD, Chipchase et al. (2012) point out that research on CPD themes and available publications in the field of physical therapy are limited. Most of the extrapolated information and data were from medical and nursing studies which were largely generalized for the field of physical therapy. An important research gap exists in the literature regarding CPD, in the field of physiotherapy as an established profession with autonomy of practice giving due consideration to the unique features far from other allied healthcare professions. Researchers such as French & Dowd (2008) argue that the lack of sufficient research has resulted in inadequate knowledge on how to address the issue of CPD in physical therapy, which in turn leads to a lack of knowledge that is essential to investigate further on the effectiveness of CPD.

2.8 Models used for evaluation of effectiveness of training

Evaluation denotes a process by which information is collected and analyzed systematically using scientific methods to ascertain the effectiveness of a program or an activity. Outcomes, are considered as the results or the benefits acquired by participants from a program. Many models have been developed to measure the training outcomes; one such popular model is the four-level training evaluation model of Kirkpatrick (1994) that is widely accepted and used across the professions. The levels are described as learner's reaction, learning, evaluation and results. Although multiple studies have used Kirkpatrick evaluation, there was always a need for a better evaluation model that can address other aspects like cost effectiveness of the training. Later on, some expanded models were developed for evaluation of training and education. For instance, the CIRO Model specifies an evaluation based on the context, input, reaction and outcome (Warr et al. 1970). Similarly, Stufflebeam et al. (2000) proposed the CIPP model which focusses on process and product in addition to context and input. Similar efforts by Phillips (1997) and Brinkerhoff (2006) to further the Kirkpatrick model resulted in a new model with a fifth level as a wider contribution.

2.9 Models used for the evaluation of the effectiveness of medical education

Continuing medical education was evaluated using the Kirkpatrick model due to the non-existence of a model that is specific for evaluating continued medical education. Researchers point out,

Table 1. Published research on continuing professional development in the field of physical therapy.

Author/Year	Type of study	Population & sample size	CPD activity and method of evaluation	Assessment conducted	Primary outcome of the study
Mays 1984	RCT	30 physical therapists	Traditional two-day course on neuro-developmental therapy, evaluated using self-report, records audit, independent observation	Soon after the activity and 6 months later	No change in practice
Bekkering et al. 2005	RCT	115 physical therapists	Active *versus* passive dissemination of information on guideline implementation for back pain evaluated using self-report	6,12,26,52 weeks after training	
Brennan et al. 2006	RCT	34 physical therapists	Two-day course + clinical improvement project on musculo-skeletal/neck pain evaluated using self-report	One year prior to CPD, one year after CPD	No improvement in Clinical outcome
Stevenson et al. 2006	RCT	30 physical therapists	Two-day course/in service training on musculo-skeletal/back pain evaluated using discharge summary questionnaire	Six months post training	No significant difference change in practice
Rebbek et al. 2006	RCT	37 physical therapists	Whiplash guideline implementation	1.5,3,6 and 12 months post activity	No patient improvement
Gunn & Goding 2008	Qualitative	11 physical therapists	CPD experience of physiotherapists using semi-structured interview		Significant change in practice and improvement in confidence and competence
Norris 2008	Longitudinal-focus group	80 physical therapists	Two-day course on back stability evaluated using questionnaire and focus group	5 years post activity	Significant change in practice
Cleland et al. 2009	RCT	19 physical therapists	Two-day course + ongoing education on musculo-skeletal/neck pain evaluated using an outcome measure as neck disability index/ NDI	One year prior to CPD and one year after CPD	No patient improvement
Li et al. 2010	Mail survey	1061 physical therapists	CPD in arthritis management evaluated using retrospective self-report		Role increased
Willet et al. 2011	RCT	43 physical therapists	Online + onsite + skill laboratory on spinal manipulation evaluated using pre- and post-participation questionnaire	Prior, 3 and 6 months post activity	Significant change in practice

however, that Kirkpatrick's evaluation model is not adequate to measure the effectiveness of CPD (French & Dowd 2008, Ahuja 2011, Chipchase et al. 2012).

There was a need for an expanded model that addresses the impact of CPD on patient care and community health. For instance, the model for measurement of outcomes of Continued Medical Education (CME) proposed by Moore et al. (2009) describes seven outcome levels as participation, satisfaction, learning, competence, performance, patient health and community health. This model suffers from lack of ease of use due to administrative constraints. The higher levels of evaluation using this model require longer duration of follow up with the participants as well as the need to have a consolidated database regarding community health. Furthermore, the studies using this model are solely restricted to medical education. For instance, Boston University School of Medicine has accepted this as the evaluation model for all their accredited continued medical education programmes. Studies using this model are currently not available, however. Existing research findings are inconclusive, regarding a validated reliable method, for evaluation of CPD across the healthcare profession.

2.10 Models used for evaluation of effectiveness of CPD in physical therapy

Generally CPD is expected to result in improved knowledge, skills and attitude of the professional. It is also expected to maintain competence and benefit the patient community by improving the quality of care. In order to understand the effectiveness of CPD, there needs to be an evaluation of the particular CPD using vigorous evaluation models.

Some scholars (Mays 1984, Brennan et al. 2006, Stevenson et al. 2006) have used one or more components of the Kirkpatrick model to evaluate the effectiveness of CPD in physical therapy. The focus of the research was, however, limited to only a few components of the model. The study conducted by Willet et al. (2011) for an evaluation of a hybrid model of CPD is the only study cited for using all the levels of this model for evaluation. The paucity of study is attributed to the administrative and methodological constraints for the implementation of higher levels of evaluations (French & Dowd 2008). Despite the lack of clear evidence about the benefits of it, professionals continue to engage in CPD, largely for compliance with statutory requirements of the profession (Landers et al. 2005).

2.11 Need for specific models for evaluation of CPD

As far as the effectiveness of CPD is concerned, existing research provides some guidance to understand the concept, although such outcomes lack clarity. The outcomes of CPD are considered far beyond mere satisfaction, knowledge, skills of the participant to a whole new dimension of comprehensive personal and development. Researchers like Stevenson et al. (2006) argue that only interactive instructional methods have been reported to improve the knowledge and skills of the participants. The review by O'Brien et al. (2001) supported the view that only interactive type of activity that can bring about changes in professional practice. In addition to that CPD is expected to bring about changes in the practice behaviour of the clinician (Nolan et al. 1995). A systematic review conducted, to measure the effectiveness of CPD in the field of dentistry supports this argument as CPD results in enriched learning and change in behaviour of the participant (Eaton et al. 2011). Currently, professional bodies are proposing a variety of evaluation methods for measuring the effectiveness of CPD. For instance, the proposed 'Portfolio model' by the Chartered Society of Physiotherapy in the UK (CSP 2013) which needs an in-depth research and is being implemented currently on a trial basis among NHS professionals.

Most organizations have adopted an input-based evaluation of CPD, which is solely based on the number of specified hours spent or credits accrued by the professional. It is clear from the available extant proposed frameworks across the professions to develop a model that focusses on output of CPD, rather than the input (WCPT 2011, CSP 2013, PSI 2010). Even though there are attempts to move beyond inputs, many of them suffer from administrative hurdles making evaluation an enduring and expensive process. Many of the suggested frameworks and models are not yet validated. Evaluation of CPD in the context of physiotherapy needs research as the current evaluation models are more suitable for evaluation of training or continued medical education. The above discussions clearly indicate serious limitations with respect to evaluation of CPD in physical therapy. It is imperative to evaluate the effectiveness of CPD using robust methodology where actual outcomes are compared against the expected outcomes.

2.12 Outcome-based evaluation of CPD

From the ongoing discussion, it is clear that CPD is recognized as an important process through which professionals can enhance their knowledge, skills and competence (French & Dowd 2008, Ahuja 2011). While most research in the field of CPD focusses on areas like identifying models of the CPD or proposing a framework for CPD rather than measuring actual outcomes (WCPT, 2006), current literature has not provided a standard taxonomy regarding the outcomes of CPD. Furthermore there is a lack of clarity in the literature regarding the use of the terms outcome, impact, results, output etc. The broader scope of CPD offers wider variety of innovative learning methods traversing from traditional lectures to informal collaborative learning in the workplace. Therefore, the outcomes of this wide variety of CPD activities will also vary. It is essential to determine what outcomes are expected from specified CPD activity while giving due consideration to the nature of the activity and/or instructional method. It should also be made clear that how the professional can ascertain that these outcome/outcomes are actually realized.

Scholars like Mays (1984), Brennan et al. (2006) and Stevenson et al. (2006) tried to identify a relationship between CPD and its effectiveness by focussing on certain outcomes such as improvement of knowledge and skills, but neglected the other outcomes. Even these limited numbers of studies cannot be generalized, focus on narrow contexts, and are grossly limited due to small sample size, methodology, and theoretical framework. WCPT has declared that CPD should provide those opportunities that result in maintaining and/or increasing the physical therapist's competence (WCPT 2011). In a qualitative study Gunn & Goding (2009) reported the evidence of change in individuals' practice and improved confidence and competence are the outcomes of CPD. Some scholars do point out, however, that it is debatable that CPD actually leads to competence and it is difficult to measure the impact of CPD in practice. It is also suggested in the literature that CPD has resulted in improvement of confidence and autonomy, enhanced competence and facilitated interchange of ideas among professionals (French & Dowd 2008). Studies addressing these aspects are, however, limited.

The number of studies investigating the effect of CPD on patient outcome is limited. It is difficult to directly relate improvement in patients' health and participation of CPD as many other factors could possibly affect the patient outcome (French & Dowd 2008). Brennan et al. (2006) and Cleland et al. (2009), however, used improvement in patient outcome as the measure for evaluating the effectiveness of CPD and the results were inconclusive. Hence the argument is that, 'are these minimum numbers of studies sufficient enough to suggest that these are the only outcomes of CPD?' It is clear that an in-depth study is paramount to identify the probable outcomes that can address this research gap regarding the CPD outcomes in the physical therapy literature.

2.13 Change in clinical practice as an outcome of CPD

CPD is regarded as a major element that has contributed to the effectiveness of physical therapists (Mays 1984, Brennan et al. 2006, Stevenson et al. 2006). The definitive goal of CPD is to ensure that the professional is capable of delivering better patient care, and change in clinical practice is one of the expected outcomes of an effective CPD (Nolan et al. 1995). Many researchers have articulated the need to link the effectiveness of physical therapists to their involvement in CPD. Studies conducted by some researchers (Mays 1984, Stevenson et al. 2006, Willet et al. 2011) focussed on the outcome of CPD as change in clinical practice in the field of physical therapy. The results are inconclusive, however, indicating that further research must be conducted to measure the effectiveness of CPD with respect to its impact in causing a positive practice change and its effect on patient outcome (French & Dowd 2008).

2.14 Change in clinical practice and evidence-based practice

Although it is evident from the literature that CPD in physiotherapy is an important component that is expected to contribute to the competence and effectiveness of physiotherapists, researchers are unable to answer how CPD contributes to evidence-based practice. Giving due consideration to the dynamic nature of the healthcare field, sometimes the professional needs to discard certain practices which were traditionally considered to be the best practice at a point in time. Based on the latest research evidence, however, those practices were found to be inadequate for the prevailing situation. Hence physical therapists need to continually update their knowledge in their practice area in order to decide whether to maintain the same practice or to adopt new evidence for the best possible patient care.

It is suggested in the literature that about 14–17 years is required for any new evidence to be accomplished in actual practice (Institute of Medicine 2009). Dissemination of the latest relevant information among professionals within the shortest possible duration can promote evidence-based care. CPD which is underpinned on the latest research is believed to be promoting evidence-based practice (Davis et al. 1999; Jull 2009). Considering the huge volume of research being published and made available through dedicated various websites for physiotherapy like PEDro, it is imperative that health professionals engage in CPD activities, to maintain up-to-date knowledge and skills, which in turn promote evidence-based quality care.

While some studies argue that CPD results in changing the professional practice behaviour of the clinician (Mays 1984, Stevenson et al. 2006, Willet et al. 2011), some others argue that CPD does not necessarily result in change in clinical practice behaviour (Chipchase et al. 2012). The number of studies in the literature addressing how CPD could lead to enhancement of evidence-based practice by bringing a change in professional practice of physiotherapists has resulted in a dilemma within the research community. Hence it is imperative to investigate further the updating role of CPD, as suggested by Sadler-Smith et al. (2000), and the role of CPD in ensuring well-trained health care professionals who actually change their practice in accordance with the latest evidence.

3 KNOWLEDGE GAP

From the foregoing literature review, it is evident that CPD as a subject is an important area that is under-researched. Current literature is not conclusive regarding the effectiveness of CPD itself, effective CPD methods, the correct combination of CPD activities, or the quantum of CPD that is needed for a professional. Furthermore there is limited literature available regarding the evaluation of CPD and the main challenge is the paucity of reliable validated models for outcome-based evaluation. Further research is required to validate the role of CPD to enhance competence or to improve clinical outcomes. From the foregoing discussion it is clear that the available studies are unable to draw definitive conclusions regarding various dimensions of CPD. This has resulted in a knowledge gap in this emerging area.

Health professionals' ability to adopt emerging evidence in clinical practice demands a positive change in practice behaviour which is critical to ensure evidence-based patient care to improve patient outcomes (WCPT 2011). The limited research has not identified how CPD can contribute to the effectiveness by change in practice behaviour of the clinician. Healthcare professionals have to prudently select the CPD activities that are efficient, convenient and can result in specific outcomes that would ultimately lead to improved patient care and community health. Stakeholders like individual professionals, professional bodies, organizations, patients and the research community also need to be provided with evidence of the effectiveness of CPD to enhance the competitiveness of the professional, which in turn benefits the patients and the community at large.

4 SUMMARY

Professionals engage in continuous professional development (CPD) activities even though there is a lack of standard outcome measurement with a definite evaluation criterion. Availability of an effective outcome-based evaluation can empower stakeholders for decision making in the selection of the most effective form of CPD. Even though the literature has recognized some of the outcomes of CPD, still there is a lack of consensus regarding the expected outcomes of various methods of CPD. Moreover there could be other outcomes which have not yet been identified. Furthermore, evidence from the literature regarding whether participation in CPD can result in specific outcomes or not is also inconclusive. Studies that have addressed the outcomes of CPD using robust methodologies are limited in the literature. In addition the lack of a validated outcome-based evaluation model for CPD itself is limiting further research for the appraisal of effectiveness of CPD.

5 CONCLUSION

CPD is acknowledged to be an ongoing process, through which professionals assume the responsibility of continuous learning through a broad range of activities intended for maintaining, developing, and enhancing knowledge & skills. CPD encompasses the overall development of the professional to improve the performance in practice. If the profession adopts a unified approach to CPD rather than a fragmented one then it offers great potential to create a well-trained and efficient workforce. Careful planning with clear implementation guidelines and an ongoing evaluation of a unified CPD system can play a critical role in having a competent healthcare workforce. This would in turn promote evidence-based practice to improve quality of care and community health at large. These arguments indicate that there is a need for a comprehensive study that addresses the effectiveness of CPD in the context of physical therapy. Further research is required for developing a better system of continuing professional development which has stronger and wider theoretical frameworks, innovative CPD instructional methods and specified evaluation criteria based on the CPD outcomes.

REFERENCES

Ahuja, D. 2011. Continuing professional development within physiotherapy: A special perspective. *Physical Therapy* 3: 4–8.
Alsop, A. 2000. Continuing professional development. In *Continuing Professional Development: A Guide for Therapists*. London: Blackwell Science.
Bekkering, G.E., van Tulder, M., Hendriks, E.J.M., Koopmanschap, M.A., Knol, D.L., Lex, M., Bouter, L.M. & Oostendorp, R.A.B. 2005. Implementation of clinical guidelines on physical therapy for patients with low back pain: Randomized trial comparing patient outcomes after a standard and active implementation strategy. *Physical Therapy* 85(6): 544–555.
Brennan, G.P., Fritz, J.M. & Hunter, S.J. 2006. Impact of continuing education interventions on clinical outcomes of patients with neck pain who received physical therapy. *Physical Therapy* 86: 1251–1262.
Brinkerhoff, R.O. 2006. *Telling Training's Story: Evaluation Made Simple, Credible and Effective*. San Francisco: Berrett-Koehler.
Chartered Society of Physiotherapy 2007. *Policy Statement on Continuing Professional Development (CPD)*. London: Chartered Society of Physiotherapy.
Chipchase, L.S., Johnston, V. & Long, P.L. 2012. Continuing professional development: The missing link. *Manual Therapy* 17(1): 89–91.
Cleland, J.A., Fritz, J.M., Brennan, G.P. & Magel, J. 2009. Does continuing education improve physical therapists' effectiveness in treating neck pain? A randomized clinical trial. *Physical Therapy* 89(1): 38–47.
Clyne. S. 1995. *Continuing Professional Development: Perspectives on CPD in practice*. London: Kogan Page.
Davis, D., Evans, M., Jadad, A. Perrier, L., Rath, D. & Sibbald G. 2003. The case for knowledge translation: Shortening the journey from evidence to effect. *British Medical Journal* 327(33): 5.

Davis, D., O'Brien, M.A., Freemantle, N., Wolf, F.M., Mazmanian, P. & Taylor-Vaisey, A. 1999. Impact of formal continuing medical education: Do conference workshops, rounds and other traditional continuing education activities change physician behavior or healthcare outcomes? *Journal of the American Medical Association* 282(9): 867–874.

Dowds, J. & French, H. 2008. Undertaking CPD in the workplace in physiotherapy. *Physiotherapy Ireland* 29: 11–19.

Dubin., S.S. 1972. Obsolescence or lifelong education: A choice for the professional. *American Psychologist* 27(2): 486–498.

Eaton, K., Brookes, J., Patel, R., Batchelor, P., Merali, F. & Narain, A. 2011. *The Impact of Continuing Professional Development in Dentistry: A Literature Review*. London: Academy of Medical Royal Colleges.

Enos, M.D., Kehrhahn, M.T. & Bell, A. 2003. Informal learning and the transfer of learning: How managers develop proficiency. *Human Resource Development Quarterly* 14(4): 369–387.

Eraut, M. 1994. *Developing Professional Knowledge and Competence*. London: Palmer Press.

French, H.P. & Dowds, J. 2008. An overview of continuing professional development in physiotherapy. *Physiotherapy* 94(3): 190–197.

Gosling, S. 1999. Physiotherapy and postgraduate study: A follow-up discussion paper. *Physiotherapy* 85(3): 117–121.

Gunn, H. & Goding, L. 2008. Continuing professional development of physiotherapists based in community primary care trusts: A qualitative study investigating perceptions, experiences and outcomes. *Physical Therapy* 95(3): 209–214.

Institute of Medicine, 2001. *Crossing the Quality Chasm: A New Health System for the 21st Century*. Washington, DC: The National Academies Press.

Institute of Medicine, 2009. *Redesigning Continuing Education in the Health Professions*. Committee on Planning a Continuing Health Professional Education Institute, Institute of Medicine. Washington, DC: The National Academies Press.

Jull, G. 2009. Invited commentary. *Physical Therapy* 89(1): 48–50.

Kirkpatrick, D.L. 1994. *Evaluating Training Programs: The Four Levels*. San Francisco: Berrett-Koehler.

Landers, M.R., McWhorter, J.W., Krum, L.L. & Glovinsky, D. 2005. Mandatory continuing education in physical therapy: Survey of physical therapists in states with and states without a mandate. *Physical Therapy* 85(9): 861–871.

Li, L.C., Hurkmans, E.J., Sayre, E.C. & Vliet Vlieland, T.P.M. 2010. Continuing professional development is associated with increasing physical therapists' roles in arthritis management in Canada and the Netherlands. *Physical Therapy* 90(4): 629–642.

Longworth, Davies, W.K. 1996. *Lifelong Learning*. London: Kogan Page.

Mansouri, M. & Lockyer, J. 2007. A meta-analysis of continuing medical education effectiveness. *Journal of Continuing Education in the Health Professions* 27(1): 6–15.

Mays, M.J. 1984. Assessing the change of practice by physical therapists after a continuing education program. *Physical Therapy* 64(1): 50–54.

Moore, D.E., Green, J.S. & Gallis, H.A. 2009. Achieving desired results and improved outcomes: Integrating planning and assessment throughout learning activities. *Journal of Continuing Education in the Health Professions* 29: 1–15.

Nolan, M., Owens, R.G. & Nolan, J. 1995. Continuing professional education: identifying the characteristics of an effective system. *Journal of Advanced Nursing* 21(3): 551–560.

Norris, C.M. 2008. Evaluation of a back stability CPD course. *Journal of Bodywork and Movement Therapies* 12(4): 305–311.

O'Brien, M.A., Freemantle, N., Oxman, A.D., Wolf, F., Davis, D.A. & Herrin, J. 2001.Continuing education meetings and workshops: effects on professional practice and health care outcomes. *Cochrane Database of Systematic Reviews* 2001(1). Art. No: CD003030.

Petty, N.J. & Morley, M. 2009. Clinical expertise: learning together through observed practice. *Manual Therapy* 14(6): 507–518.

Pharmaceutical Society of Ireland (PSI) 2010. *Review of International CPD Models. Final Report*. Dublin: PSI.

Phillips, J.J. 1997. *Return on Investment in Training and Performance Improvement Programs*. Boston: Butterworth-Heinemann.

Physiotherapy Board of Australia (PBA) 2010. *Continuing Professional Development Registration Standard*. Canberra: Physiotherapy Board of Australia.

Rebbek, T., Maher, C.G. & Refshauge, K.M. 2006. Evaluating two implementation strategies for whiplash guidelines in physiotherapy: A cluster randomized trial. *Australian Journal of Physiotherapy* 52(3): 165–174.

Sadler-Smith, E., Allinson, C.W. & Hayes, J. 2000. Learning preferences and cognitive style: Some implications for continuing professional development. *Management Learning* 31(2): 239–256.

Stein, A.M. 1998. History of continuing nursing education in the United States. *Journal of Continuing Education in Nursing* 29(6): 245–252.

Stevenson, K., Lewis, M. & Hay, E. 2006. Does physiotherapy management of low back pain change as a result of an evidence-based educational programme? *Journal of Evaluation in Clinical Practice* 12(3): 365–375

Stufflebeam, D.L., Madaus, G.F. & Kellaghan, T. 2000. *Evaluation Models: Viewpoints on Educational and Human Services Evaluation.* Boston: Kluwer Academic Publishers.

Umble, K.E. & Cervero, R.M. 1996. Impact studies in continuing education for health professionals: A critique of the research syntheses. *Evaluation and the Health Profession* 19(2): 148–174.

Warr, P., Bird, M. & Rackham, N. 1970. *Evaluation of Management Training.* London: Gower Press.

Willet, G.M., Johnson, G.C. & Jones, K. 2011. The effect of a hybrid education course on outpatient physical therapy for individuals with low back pain. *Internet Journal of Allied Health Sciences and Practice* 9(1).

World Confederation of Physical Therapy (WCPT) 2011. *WCPT Guideline for Delivering Quality Continuing Professional Development for Physical Therapists.* London: WCPT.

World Federation for Medical Education (WFME) 2003. Continuing Professional Development (CPD) of Medical Doctors. *WFME Global Standards for Quality Improvement.* Denmark: WFME Office, University of Copenhagen.

Impact of quality assurance on accountability in policy networks

Maitham Al Oraibi & Samia Costandi
Ahlia University, Kingdom of Bahrain

Tillal Eldabi
Brunel University, UK

ABSTRACT: This paper explores the research question of how publicly reported quality assurance (QA) review results of public service providers can impact the accountability in policy-making networks that have been created with the scope of reforming such public service provision. It does that by combining different theoretical concepts of two different perspectives in public administration: new public management and networked governance. The research uses empirical data collected from semi-structured qualitative interviews to construct a working conceptual model. Results of thematic analysis suggest that publicly reported QA is expected to have a positive impact on accountability through six various mechanisms: creating an accountability environment, getting access to information on performance, promoting self-accountability, having more control, balancing of power, and managing expectations. The emergent developing focal theory will then be used to collect more data to validate the constructed key and sub-propositions.

1 INTRODUCTION

New public management (NPM) and networked governance are considered to be two forms of "third-party" or alternative governance (Heinrich 2009, Rethemeyer & Hatmaker 2007). While NPM translates ideas from the private sector and market mechanism, the network approach focusses on policy making through inter-organizational co-ordination (Klijn & Koppenjan 2000). The main concept of this third-party governance is the devolution of power and roles in the provision of goods and services to the public from the central government to these third-party, non-governmental entities (Stoker 1998).

NPM policies started in the 1980s, driven by cost efficiency and neo-liberal perspectives on reforming public service provision. The main concepts of the NPM policies evolve around decentralizing management, and making better use of market and competition incentives in the provision of public services (UNRISD 1999, Ferlie et al. 2005). Along with NPM policies came a big trend towards a broader but rapidly evolving "performance measurement society" to cover quality assurance and inspection activities (Power 1997, 2000, 2003, Justesen & Skaerbaek 2010, Bowerman et al. 2000, Maijoor 2000). Performance measurement came to address the "limits of privatization" and inadequate monitoring agencies (Amirkhanyan 2008). Its impact, however, on improving the outcomes of a public programme is still debatable (Propper & Wilson 2003, Marshall et al. 2003, Werner & Asch 2005, Hibbard et al. 2005, Robinowitz & Dudley 2006, Lindenauer et al. 2007, Hodgson et al. 2007, Lim 2009). Part of the dilemma in evaluating the effectiveness of performance measurement and reporting is its ability to reconcile varying needs and expectations of stakeholders (Humphrey & Owen 2000, Cotton et al. 2000, Lansky 2002, Bolton & Hyland 2003, Preston & Hammond 2003; Klessig et al. 2000, Skinner et al. 2004, Joseph & Joseph 1997). This is one of the reasons why countries opt for a networked or collaborative governance mode.

Network is a term that typically refers to "multi-organizational arrangements for solving problems that cannot be achieved, or cannot be achieved easily by single organization" (Agranoff & McGuire 2001, p. 296). Networks can have mandatory or voluntary participation, involving public or private actors (Meier & O'Toole 2001). The term 'network' is used along with other terms such as partnership, alliance, collaboration, co-ordination, co-operation, joint working and multi-party or inter-organizational working, and it is also used in distributed leadership (Huxham 2003, Ospina & Saz-Carranza 2010, Martin et al. 2008, Gazley 2010).

The literature suggests that there is a strong move in governance from hierarchical or command-and-control mechanisms to more collaborative governance that includes networks of government, both for-profit and non-profit actors (Silvia 2011, Huxham 2003, Isett et al. 2011, Martin et al. 2008, Bailey, 2003). One of the key challenges that networks can face, however, is performance measurement of public services delivery or policy implementation (Lambright et al. 2010). Whilst working in networked or collaborative settings can bring forward some advantages over hierarchical governance (Lambright et al. 2010), managing such networks have particular challenges (Agranoff & McGuire 2001). The challenges reported in the literature can be either structural, pertaining to the structural complexity of the network setting (Huxham et al. 2000, Acar et al. 2008, Acar & Robertson 2004), or behavioural, emanating from the behaviour of network members within a network (Acar & Robertson 2004, O'Connell 2005). On top of these challenges is accountability, which warrants special management strategies to manage (Agranoff & McGuire 2001, Klijn & Koppenjan 2000). The challenges of accountability reported in the literature vary considerably according to the type of the inter-organizational network.

The definitions of accountability are closely linked to "answerability" for performance, "to whom" and "to what" elements (Dicke 2002, Romzek 2000). Answerability is usually to a higher authority in a bureaucracy or inter-organizational chain of command (Dicke 2002). The three elements, answerability, to whom and to what, are part of the challenge of accountability in a network setting, compared to a vertical bureaucratic organization.

In some contexts, however, tools of the two modes of governance can be used together to reform public services. Networked governance is coupled with other initiatives such as independent publicly reported quality assurance (referred to hereinafter as QA), whereby quality of such public services is continuously reviewed, measured and reported in the public domain. These reports not only measure the performance of individual service providers directly, but also give an aggregate picture of the performance of the sector, for which the network, or some of its members, are mandated to regulate and improve its overall performance. In the literature, accountability in networks is well covered from various perspectives. There is, however, no theoretical framework that underpins conceptualizing of the impact of publicly reported QA on the accountability in such context of policy-making networks. One possible explanation for this gap in the literature is that the two aspects, quality inspection and networked governance, come from two different paradigms in the public administration arena: new public management governance (where quality inspection perspective comes from) and networked or collaborative governance (from which networks perspective comes). This research aims at developing such a conceptual model. The research tries to answer this question: 'How can publicly reported QA of public service-providers impact accountability in policy-making networks scoped to reform the quality of such public services?'

The paper is divided into four sections: in the first section, relevant literature is reviewed for identifying possible, or expected, ways in which QA reports can impact accountability in such networks. An initial key proposition and/or background theory is developed at this stage. The second section explains the method of the research, and how the empirical data are collected and analysed. The third section discusses in more detail the collected qualitative data, builds a set of initial, but specific, sub-proposition that can explain the anticipated impact of QA reports on accountability. The initial key proposition is then updated and a more extended proposition and sub-propositions, or focal theory, is constructed accordingly. The paper concludes by explaining the outcomes of the research, and the steps needed to validate the conceptual model developed here.

2 LITERATURE REVIEW: BACKGROUND THEORY

This section reviews the findings from published literature on accountability in network settings. The published research treats accountability in various contexts, none of which considers the existence of publicly reported performance measurement such as quality inspection. Realizing this gap in the literature, the next discussion aims at constructing an initial and generic proposition, background theory, on how publicly reported QA can impact accountability in such collaborative policy making networks, based on a number of expected ways, from literature, in which such a proposed impact can materialize. The expected themes for this effect are highlighted in the following sections.

There are different types in which accountability works depending on the structure and the degree of control or authority exercised by parties. The most common types of accountability are the hierarchical, legal and political accountability (Aucoin & Heintzman 2000, Page 2004, Romzek 2000). In addition to these types, two more types of accountability are discussed in the literature that could explain how accountability might work in settings where none of the hierarchical, legal or political mechanisms exist such as network settings. O'Connell (2005, p. 92) uses the term "accountability environment" to refer to collective accountability that comes form "the interactions of multiple parties [and which] appears to be characteristic of many government programs". This definition encompasses other terms that are used to define similar accountability such as "professional accountability" (Page 2004, Romzek 2000), "accountable culture" (Dubnick & Frederickson 2009), and mutual or dialogue-based accountability. According to this definition, one might expect that by public reporting of QA results, the surrounding environment around the key players in charge of the quality of services, for example, service providers and network members, will be more motivated to apply various mechanisms available at their discretion to hold these players accountable for improving the quality of services.

Acar et al. (2008) list five functions of accountability in network settings, the first of which is mapping and manifesting expectations of stakeholders from the network or partnership. Page (2004) lists "managing expectations" as an example of an internal platform in which accountability functions in a network. By public reporting of QA results, it is expected that public reports may help in shaping expectations of either internal members of a network or external stakeholders. This can be yet another possible mechanism in which publicly reported QA impacts accountability in a network setting.

Accountability in network settings can serve different objectives. One of the key concepts in NPM is the accountability for results (Bardach & Lesser 1996) or "management for results" (Page 2004). Aucoin & Heintzman (2000) identify threefold objectives of accountability in managing governmentally funded programmes: control, assurance, and continuous improvement. The three objectives overlap in several ways, but they all aim at ensuring and maximizing efficiency in the use of public money. QA by public reporting is expected to motivate stakeholders to apply more controls in order to improve quality of public services, as well as encourage decision makers, regulators, and funding agencies to adopt more of a "management for results" approach, either to hold the service providers more accountable, or to improve their positions in front of parties to whom they are answerable in a network.

"Access to information" on performance is listed on top of the challenges that face networks (Acar et al. 2008, Acar & Robertson 2004, Dubnick 2005, Page 2004, Romzek, 2000), whereas the main motive behind NPM policies is the demand for greater accountability and transparency (Justesen & Skaerbaek 2010). At this point, one can raise the following question: 'Will an independent quality assurance and reporting help in overcoming the challenge of access to performance data, especially if the reports are made public to all stakeholders?' From a theoretical stance, the answer will be quite possible.

Related strongly to the challenge of access to information is another challenge of "measurability of performance". According to Acar & Robertson (2004), measurability is at the heart of constraints to access of information that faces accountability. Page (2004) suggests that in order to be accountable for results, networks should have the capacity to measure performance in the first place. Romzek (2000) stresses on the importance of measurability especially for government reform types of projects. QA serves directly to measure and report performance of providers, and hence collectively the sector. It is expected, therefore, that public reports of QA help in providing the required measurability of performance to strengthen the accountability in a network.

One of the challenges in network due to the hierarchical authority arrangement is the distribution of power inside a network (Agranoff 2006, Huxham 2003), where "one part has no hierarchical authority over its partners and no full control over the performance" (Acar et al. (2008, p. 4). It is recognized that asymmetry of power, in terms of amount and type of power that members of a network hold, is one of the main challenges that may lead to major implications on accountability (Acar & Robertson 2004, Crosby & Bryson 2005, Ansell & Gash 2007). The imbalance in collaboration can be caused by real or perceived differences in power occurring at organizational or individual levels (Huxham et al. 2000). By public reporting of quality inspection, and by putting all the stakeholders on a balanced ground of accountability, will QA reports help in "balancing the power" by diminishing the imbalance that usually exists within a network? That is the expected theme here.

Based on the aforementioned discussion, and the number of expected ways in which publicly reported QA can impact accountability of both, service providers for providing expected quality services, and network members for achieving the common objectives of a network in reforming or improving the quality of such public services, the following overall proposition has been put together:

Key Proposition (initial): 'By making its reports on performance of service providers public to all stakeholders, independent quality assurance can impact the accountability in a network.'

3 DATA COLLECTION AND ANALYSIS

The research question here is of an explanatory type (Blaikie 2010), seeking an explanation of the association between QA public reporting and accountability in policy-making networks. Maxwell (2013) explains this as a causal explanation, in which the researcher uses a qualitative approach to "ask how (x) plays a role in causing (y)".

In the literature, there is no conceptual framework that depicts the impact of publicly reported QA on accountability in network settings, but there is an abundance of papers that discuss different aspects of accountability in networks or inter-organizational collaborative settings. The objective of this research is to build a conceptual model that explains the impact of publicly reported QA on accountability in a network. Hence, this research uses mixed deductive-induction approach (Miles et al. 2014, Boardman 2011). The literature review outcomes were used first to build an initial proposition or background theory. Open-ended semi-structured interview protocol was used to collect data. The open-ended question was on how the QA public reports might have impacted accountability in any way (positive or negative). The data then undergo process coding (Miles et al. 2014) to reach to second-order themes and build a number of sub-propositions (Saunders et al. 2009), or "working hypothesis" as termed by Miles et al. (2014), which were then used to construct an initial model of a "focal theory". This model is developing and still needs to be supported by more interviews.

The empirical context of this research is policy-making networks related to the Education Reform programme in the Kingdom of Bahrain. In 2007, Bahrain embarked on a major reform programme to enhance the quality of basic education, vocational education and training (VET) and higher education (HE). As part of this programme, an independent quality assurance authority was established to review and report to the public domain the quality of all private and public education and training providers on the island. The review is cyclic and covers all schools, VET and HE intuitions. Concurrently, a number of committees, or networks, comprising representatives of various stakeholders from the private and public sectors were formed to review and make necessary policies and strategies to improve the overall quality of education and training sectors.

For this research, four experts were interviewed (Saunders et al. 2009) using the open-ended semi-structured interview protocol. The four interviewees were selected to represent the whole spectrum of the stakeholders in these networks as follows:

- Interviewee 1: Represents quality inspectors and sits on two committees related to VET.
- Interviewee 2: Represents a regulatory body and sits on three committees related to basic schooling, VET and HE.

- Interviewee 3: Represents a HE provider and sits on a number of committees related to HE.
- Interviewee 4: Represents a funding agency and sits on a committee related to VET.

Interviews were audio taped, transcribed, analysed, coded using process coding approach (Miles et al. 2014) and analysed using thematic analysis over two stages: first-order and second-order themes (Attride-Stirling 2001, Sobh & Perry 2006). An Excel sheet was then used to list all the codes and filter them in patterns. Accordingly, the second-order themes were identified (refer to Table 1). The results of this stage of data analyses were then used to revise the initial key propositions, construct six (working) sub-propositions, and to build the initial, developing, focal theory (Figure 1).

4 RESULTS AND DISCUSSION: FOCAL THEORY

Responses from interviewees were coded and analysed to identify six second-order themes. The theme about the balancing of power still needs to be supported by further data as it was not strongly evident. In addition, one new theme emerged from the data: this theme is about promoting self-accountability. This theme has no relevance to any of the expected themes identified during the literature search that was reviewed in building the background theory. The following discusses in detail the development of the second-order themes, and, hence, sub-propositions.

4.1 *Creating accountability environment*

This theme represents the various ways in which public reporting of QA can help in creating or increasing the effects on the accountability of either member organizations within a network or service providers. The accountability environment here represents accountability methods outside the formal ones, that is, hierarchical and legal types (Aucoin & Heintzman 2000, Page 2004, Romzek 2000). This comprises aspects that can lead to having generally more demanding and questioning stakeholders around individual organizations or service providers.

The four interviewees agreed that the QA initiative has lead to an accountability environment, in one way or another. The reported codes by the interviewees can be of mechanisms that create an accountability environment surrounding service providers, members within a network, or both:

- The general public becomes more concerned about the quality and performance of services, using QA results to inform their decisions in selecting service providers, being more proactive in providing feedback to regulators on the performance of providers.
- Government organizations become more concerned about public money, exercise more care in select providers and awarding contracts or funds.
- Employers, parents and learners, act as end customers of service providers, become more selective in choosing a provider to pay for required services.
- Pressure exercised on member organizations of a network by their counterparts from within a network in reviewing QA or agreed-upon key performance indicators that are based on QA reports.
- Pressure is faced by individual members of a network from their superiors in their organizations once the reports are made public.

The quote below illustrates an example of such accountability in the surrounding environment and its impact on the service provider. The words in double inverted commas are the key words used to identify the first order themes and their associations and impact:

'Well it is simple, [institute] doing bad "accountable" to improve "in front of the parents", of the "employers" sending people to training institutes, "accountable to [funding organizations]" sending many beneficiaries, accountable to the "government" sending students to training institutes' (Interviewee 4).

The members of a network also receive such a pressure from the environment. The following quote illustrates such a pressure from within a network:

'We "go in-depth in the report" within these committees and there is even embarrassing, sometime, to [members] either by the head of the committee, or by other [members], everybody "accepting" the outcome or the result of the QQA.' (Interviewee 2).

Pressure can also be exerted on members from their superiors in their own organizations, as in this quote, for example:

'Yes, both of these Ministries are "under big telescope" of the higher management within the government and "members that come to these meetings", they are "accountable in front of their minister within their ministry" and they follow up with them if there are bunch of reports that are published periodically." (Interviewee 1).

Please note that some of these mechanisms, reported above, can develop further into other forms of accountability. For example, the concern and selectivity of governmental or funding organizations can be reflected, once the provider was selected and awarded contract/fund, into more formal contractual or legal accountability (Aucoin & Heintzman 2000, Page 2004, Romzek 2000). By the same token, pressure from superiors within organization members can be developed further into hierarchical accountability. This theme, however, considers the general and less formal mechanisms of accountability that the surrounding environment exercises on providers and network members in working together to improve the overall quality of such public services.

The above discussion demonstrates or reflects the idea that the publicly reported QA supports the following sub-proposition:

P1.a: 'Publicly reported QA helps creating accountable environment around service providers and members of a network, who are working towards improving quality of related public service.'

4.2 *Getting access to information on performance*

One of the key challenges to accountability is the access to reliable and credible information on the performance of sectors, networks and individual service providers (Acar et al. 2008, Acar & Robertson 2004, Dubnick 2005, Page 2004, Romzek 2000). It was expected before the interviews that the responses of interviewees would converge into three themes: accesses to information about performance, measurability of performance, and management for results.

The responses of the four interviewees support strongly the link between QA reports and getting such reliable, credible and independent information about performance, which can be then used as bases for accountability, concomitantly, this information can be used in managing the sector, networks, or individual service providers. The answers of the interviewees, however, make little distinction between the three expected themes prior to conducting the interviews. One possible reason for this is that what matters for the network members is the availability of such data. The usage of the data comes by default, as a result rather. This quote explains this:

'Getting more information, well, it is questionable, "the right information" I would say, rather than more information so that what "really matters".' (Interviewee 4)

The four interviewees indicate that QA reports can fulfil the requirement of this information to have better accountability within a network. The responses, however, differ in the way they perceive how the information could then be used by the various members and stakeholders. As an example of this, the following is a quote from one interviewee:

'I think it "put very important solid criteria into the evaluation", in the past we used to look at the management team, the lecturer and the content, but now the "quality assurance report you covers a lot".' (Interviewee 4)

The following quote suggests how the participants make another use of such information in monitoring the performance of a network:

'When we start the [network], we put "KPIs for the committee", and one of the KPIs is how the grading, the "outcome of the reviews", are happening over time and whether those institutes who failed are improving or not.' (Interviewee 1)

Here, the first-order themes include aspects related to obtaining the right information about performance that can be used to inform funding decisions, monitoring the performance of providers, updating legislative tools, and the monitoring and management of the performance of the network and sector as a whole. The following is the list of first-order themes that were gauged from the interviews:

- Obtaining consistent, credible and independent measurement of performance of providers.
- Providing reliable information that can be used for funding decisions.
- Providing information that can identify the responsibilities of each member within a network to improve performance.
- Providing information on performance that can be used to update legislative tools.
- Providing information that can be used to monitor the performance of a network.

The above themes dovetail into one general theme, which is "getting access to information on performance". The quotes from the interviewees suggest a strong link between QA reports and its function in providing such information on performance with the required attributes of credibility, independence and reliability. The next sub-proposition is then constructed based on this:

P1.b: 'Publicly reported QA provides required information on performance that can be used by stakeholders for their own management decision-making.'

4.3 Managing expectations

Managing stakeholders' expectations is one way of how accountability works in a network (Page 2004, Acar et al. 2008). Based on this theoretical dimension, it was expected that by making QA reports available in the public domain, QA would help in raising the expectations of stakeholders and the general public to improving the standards of quality of public services.

Two of the interviewees reported such a link between QA reports and their impact on managing expectations of parents for better quality of schools, HE, and vocational institutes, and creating a general public demand for the improvement of quality. The following are two quotes that support this theme:

'On a social level, there is something been noted. If you want to "send your kids to a certain school, private school or public school, or a university", you don't know what to do. Today, as a family, "everybody" will say "go to the website of the QQA".' (Interviewee 2).

'I think there is ongoing "national pressure" on things to get better and the QQA is part of that national pressure. I mean the "people of Bahrain understand better" now that the quality assurance and assuring the quality of education and training in this country and improving them is a must ... I think all the members of these various committees will "have that kind of accountability" to continue the internal and external "improvements".' (Interviewee 1).

This theme comes from two interviews. The first-order themes that comprise this are: raising the expectations of parents and families in regard to the quality of schools, HE and vocational institutes; and, creating a public demand for continuous improvement of those public services. Although the collected quotes refer to parents and the general public only, one could expect that the same trend would be extrapolated to include other members of a network and relevant stakeholders (for example, employers, regulators and funding organizations). These empirical data can be used at this stage to formulate an initial sub-proposition as follows:

P1.c: 'Publicly reported QA can help in managing expectations of members, stakeholders and the general public regarding having better quality of public services.'

4.4 Promoting self-accountability

This is a new theme that emerges as a separate pattern from the responses of the interviewees. It was not initially expected. Two interviewees suggested that public reporting of QA promotes more effective and responsive self-review and accountability within service providers, whether in response to QA reports about their performance or performance of other providers as well.

One interviewee, representing a Higher Education Institute in a network, describes their initiative in self-review and its link to the issue of accountability as follows:

'What [we] decided to do, and I think it is "very innovative", of course I am biased about it, but was we decided that we would "create our self review system" ... so we developed our own system with our own criteria and we embedded these other criteria within it, so how this "linked to your question which is about the public accountability".' (Interviewee 3).

Self-accountability becomes more serious when funding decision is linked with public reporting, as explained in this quote:

'I think after public reports, the reputation of training provider became on the edge, "either to improve and change according to the recommendation of the report, or [providers] will not only lose the government funding, but even the private or the public funding".' (Interviewee 2).

The first-order themes that comprise this theme include: serious adoption of internal quality systems, implementing self-review procedures within providers, self-diagnoses/analyses against QA review indicators and criteria, more emphasis by providers on the quality of services (for example, teaching and learning), encouraging other providers to improve by reading QA reports of other providers.

The codes aggregated for this theme suggest a positive link between the public reporting of QA results and promoting self-accountability within service providers. By the same token, one would expect to find the same pattern, with less seriousness perhaps, within regulators or government organizations in charge of overall quality of education and training sectors. This extension needs further empirical data to support it. The following sub-proposition can be formed then:

P1.d: 'Publicly reported QA promotes practices of self-review and accountability within service providers and regulators.'

4.5 Balancing of power

Although balancing of power is reported in the literature as one of the challenges and a key strategy in managing accountability in a network (Acar et al. 2008, Acar & Robertson 2004, Crosby & Bryson 2005, Ansell & Gash 2007, Huxham 2003), and although this theme was expected to emerge here, the responses of the interviewees do not support this theme strongly. It was reported by one interviewee in the following quote:

'It is basically "more accountable" to be in "power balance" which has been created by the report itself. Earlier when you take a decision, nobody was asking based on what you took your decision. Today, you have a "foundation", which is basically the report.' (Interviewee 2)

At this stage it is a premature to predict, one can form an expected sub-proposition when more supporting evidence has been collected:

P1.e: 'Publicly reported QA helps in the balancing of power that is needed for better accountability in a network.'

4.6 Having more control

This theme represents a trend of actions or measures that regulatory and funding agencies undertake to have more control over service providers, either in direct response to QA reports, or as a result of the various pressures from the above-mentioned mechanisms, such as the accountability environment or managing expectations of stakeholders. This happens to be in line with what was expected previously as an identified trend from the literature (Aucoin & Heintzman 2000).

Indeed, the responses from interviewees support this theme strongly. All of the interviewees have cited some examples to support this. The following quotes clearly depict this trend:

'I think now that there "has been public" ... in the newspaper they have "closed universities, they have stopped programs", I think that "the sector as a whole can see that".' The interviewee then adds further 'the [regulator] do 'have teeth and they really will do something' that you cannot just run something sloppy. (Interviewee 3)

Table 1. Two-stage thematic analysis results.

First-order themes	Second-order theme
General public being more concerned, using QA results to inform decisions of selecting service providers, being more proactive in providing feedback to regulators on performance of providers, government organizations becoming more concerned about public money, exercising more care in selecting providers and awarding contracts/funds, employers, parents and learners (end customers) being more selective in choosing provider, pressure faced by member organizations of a network from counterparts, pressure faced by individual members of a network from their superiors in their organization.	Creating accountability environment
Getting consistent, credible and independent measurement of performance of providers, providing reliable information, providing information to identify responsibilities of each member within a network, providing information on performance to update legislative tools, providing information to monitor the performance of a network.	Getting access to information on performance
Raising expectation of parents and families with regards to the quality of schools, HE and vocational institutes; and creating a public demand for continuous improvement of public services.	Managing expectations
Serious adoption of internal quality systems, implementing self-review procedures within providers, self-diagnoses/analyses against QA review indicators and criteria, more emphasis by providers on the quality of services (for example, teaching and learning); and encouraging other providers to improve by reading others' QA reports.	Promoting self-accountability
One code only here refers to the use of reports to justify/challenge decisions of members in a network.	Balancing of power
Regulatory actions based on reports, linking funding with outcomes of reports, implementing more stringent monitoring strategies on licensed providers, updating regulations in response to the outcomes of reports, having sticks/carrots to improve performance.	Having more control

In the same way, regulatory organizations also respond to the QA reports as in the following quote:

'We found out that there are lot of institutes who were not the right institute to be "provide training". So, we "decided to follow up heavy procedure" with them based on the law, which is "in our favour" to shut them down if they are not improving.' (Interviewee 2)

First-order themes found in this category are as follows:

- Regulatory actions based on public reporting.
- Linking funding with outcomes of reports.
- Pressure on regulators to implement more stringent monitoring strategies on licensed providers.
- Updating existing regulations in response to the outcomes of reports, to have sticks/carrots to improve performance.

There are many quotes in this theme that suggest a very positive link between QA public reporting and tightening control measures on providers to improve further. This takes us to the next sub-proposition:

P1.f: 'Publicly reported QA leads to tightening control measures on service providers to improve further.'

A summary is given in Table 1.

Based on the discussions above, and the sub-propositions constructed, the next model depicts the proposed (developing) focal theory for the impact of QA reporting on accountability (Figure 1). At this stage, the model is still a working model, and needs to be further validated with more empirical data.

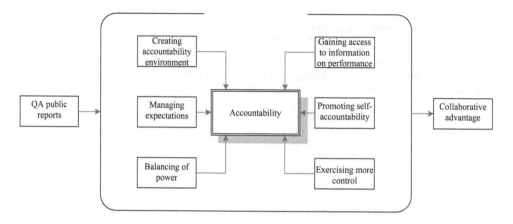

Figure 1. Developing focal theory (developed for this research).
*The collaborative advantage aspect is not within the scope of this paper, it will be discussed in more detail in a subsequent paper being developed by the authors.

5 CONCLUSIONS

Thematic analyses of the responses of experts' interviews suggest that publicly reported QA can have a positive impact on quality. In conclusion, the initial proposition which was developed based on the literature review was then revised as follows:

Key Proposition (revised): 'By making its reports on performance of service-providers public to all stakeholders, independent quality assurance can have a positive impact on the accountability in policy-making networks.'

There are six sub-propositions that support this conclusion:

P1.a: 'Publicly reported QA helps creating accountable environment around service providers and members of a network, who are working towards improving quality of related public service.'
P1.b: 'Publicly reported QA provides required information on performance that can be used by stakeholders for their own management decision-making.'
P1.c: 'Publicly reported QA can help in managing expectations of members, stakeholders and the general public regarding having better quality of public services.'
P1.d: 'Publicly reported QA promotes practices of self-review and accountability within service providers and regulators.'
P1.e: 'Publicly reported QA helps in the balancing of power that is needed for better accountability in a network.'
P1.f: 'Publicly reported QA leads to tightening control measures on service providers to improve further.'

The revised key proposition and the six associated sub-propositions are used to construct a working model depicting the subject impact (Figure 1). The emerging focal theory (key proposition and sub-propositions) will be used as a basis for the collection of wider data from a multi-case study to fully hypothesize the impact. This research has not yet looked at the outcome of such an impact, either on individual network members, or on networks, or at the sector levels. This will be dealt with separately as part of "collaborative advantage" research in this context.

REFERENCES

Acar, M. & Robertson, P. 2004. Accountability challenges in networks and partnerships: Evidence from educational partnerships in the United States. *International Review of Administrative Sciences* 70: 331–344.

Acar, M., Guo, C. & Yang, K. 2008. Accountability when hierarchical authority is absent: Views from public-private partenership practitioners. *The American Review of Public Administration* 38(1): 3–23.

Agranoff, R. 2006. Inside collaborative networks: Ten lessons for public managers. *Public Administration Review* Special Issue: 56–65.

Agranoff, R. & McGuire, M. 2001. Big questions in public network management research. *Journal of Public Administration Research and Theory* 3: 295–326.

Amirkhanyan, A. 2008. Collaborative performance measurement: Examining and explaining the prevalence of collaboration in state and local government contracts. *Journal of Public Administration Research and Theory* 19: 523–554.

Ansell, C. & Gash, A. 2007. Collaborative governance in theory and practice. *Journal of Public Administration Research and Theory* 18: 543–571.

Attride-Stirling, J. 2001. Thematic networks: An analytic tool for qualitative research. *Qualitative Research* 1: 385–405.

Aucoin, P. & Heintzman, R. 2000. The dialectics of accountability for performance in public management reform. *International Review of Administrative Sciences* 66: 45–55.

Bailey N. 2003. Local strategic partnerships in England: The continuing search for collaborative advantage, leadership and strategy in urban governance. *Planning Theory & Practice* 4(4): 443–457.

Bardach, E. & Lesser, C. 1996. Accountability in human Services: Collaboratives for what? and to whom? *Journal of Public Administration Research and Theory* 6(2): 197–224.

Blaikie, N. 2010. *Designing Social Research*, 2nd ed., Polity Press.

Boardman, C. 2011. Organizational capital in boundary-spanning collaborations: Internal and external approaches to organizational structure and personnel authority. *Journal of Public Administration Research and Theory* 22: 497–526.

Bolton, T. & Hyland, T. 2003. Implementing key skills in further education: Perceptions and issues. *Journal of Further and Higher Education* 27(1): 15–26.

Bowerman, M., Raby, H. & Humphrey, C. 2000. In search of the audit society: Some evidence from health care, police and schools. *International Journal of Auditing* 4: 71–100.

Cotton, P., Fraser, I. & Hill, W. 2000. The social audit agenda: Primary health care in a stakeholder society. *International Journal of Auditing* 4: 3–28.

Crosby, B. & Bryson, J. 2005. A leadership framework for cross-sector collaboration. *Public Management Review* 7(2): 177–201.

Dicke, L. 2002. Ensuring accountability in human services contracting: Can stewardship theory fill the bill? *The American Review of Public Administration* 32: 455–470.

Dubnick, M. 2005. Accountability and the promise of performance: In search of the mechanisms. *Public Performance & Management Review* 28(3): 376–417.

Dubnick, M. & Frederickson, H. 2009. Accountable agents: Federal performance measurement and third-party government. *The Journal of Public Administration Research and Theory* 20: 143–159.

Ferlie, E., Lynn, L., Jr & Pollitt, C. 2005. *The Oxford Handbook of Public Management*. Oxford: Oxford University Press.

Gazley, B. 2010. Linking collaborative capacity to performance measurement in government – non-profit partnerships. *Nonprofit and Voluntary Sector Quarterly* 39(4): 653–673.

Heinrich, C. 2009. Third-party governance under no child left behind: Accountability and performance management challenges. *Journal of Public Administration Research and Theory* 20: i59–i80.

Hibbard, J., Stockard, J. & Tusler, M. 2005. Hospital performance reports: Impact on quality, market share and reputation. *Health Affairs* 4(24): 1150–1160.

Hodgson, A., Edward, S. & Gregson, M. 2007. Riding the waves of policy? The case of basic skills in adult and community learning in England. *Journal of Vocational Education & Training* 59(2): 213–229.

Humphrey, C. & Owen, D. 2000. Debating the 'power' of Audit. *International Journal of Auditing* 4: 29–50.

Huxham, C. 2003. Theorizing collaboration practice. *Public Management Review* 5(3): 401–423.

Huxham, C., Vangen, S. & Eden, C. 2000. The challenge of collaborative governance. *Public Management: An International Journal of Research and Theory* 2(3): 337–358.

Isett, K., Mergel, I., LeRoux, K., Mischen, P. & Rethemeyer, R. 2011. Networks in public administration scholarship: Understanding where we are and where we need to go. *Journal of Public Administration Research and Theory* 21: i157–i173.

Joseph, M. & Joseph, B. 1997. Service quality in education: A student perspective. *Quality Assurance in Education* 5(1): 15–21.

Justesen, L. & Skaerbaek, P. 2010. Performance auditing and the narrating of a new auditee identity. *Financial Accountability and Management* 26(3): 325–342.

Klessig, J., Wolfsthal, S., Levine M., Stickley, W., Bing-You, R., Lansdale, T. & Battinelli, D. 2000. A pilot survey to define quality in residency education. *Academic Medicine* 1(75): 71–73.

Klijn, E. & Koppenjan, J. 2000. Public management and policy networks. *Public Management: An International Journal of Research and Theory* 2(2): 135–158.

Lambright, K., Mischen, P. & Laramee, C. 2010. Building trust in public and non-profit networks: Personal, dyadic, and third-party influences. *The American Review of Public Administration* 40(1): 64–82.

Lansky, D. 2002. Improving quality through public disclosure of performance information. *Health Affairs*, 4(21): 52–62.

Lim, D. 2009. Testing the effectiveness of a quality assurance system: The example of Hong Kong. *Journal of Vocational Education and Training* 61(2): 183–202.

Lindenauer, P., Remus, D., Roman, S., Rothberg, M., Benjamin, E., Ma, A. & Bratzler, D. 2007. Public reporting and pay for performance in hospital quality improvement. *The New England Journal of Medicine* 5(356): 486–496.

Maijoor, S. 2000. The internal control explosion. *International Journal of Auditing* 4: 101–109.

Marshall, M., Shekelle, P., Davies, H. & Smith, P. 2003. Public reporting on quality in the United States and the United Kingdom. *Health Affairs* 3(22): 134–148.

Martin, G., Currie, G. & Finn, R. 2008. Leadership, service reform and public-service networks: The case of cancer-genetics pilots in the English NHS. *Journal of Public Administration Research and Theory* 19: 769–794.

Maxwell, J. 2013. *Qualitative Research Design: An Interactive Approach*, 3rd ed. New York: Sage Publications.

Meier, K. & O'Toole, L., Jr. 2001. Managerial strategies and behavior in networks: A model with evidence from U.S. public education. *Journal of Public Administration Research and Theory* 3: 271–293.

Miles, M., Huberman, A. & Saldana J. 2014. *Qualitative Data Analysis: A Methods Sourcebook*, 3rd ed. New York: Sage Publications.

O'Connell, L. 2005. Program accountability as an emergent property: The role of stakeholders in a program's field. *Public Administration Review* 65: 85–93.

Ospina, S. & Saz-Carranza, A. 2010. The behavioural dimension of governing inter-organizational goal-directed network – managing the unity-diversity tension. *Journal of Public Administration Research and Theory* 21: 327–365.

Page, S. 2004. Measuring accountability for results in interagency collaboratives. *Public Administration Review* 64: 591–606.

Power, M. 1997. *The Audit Society: Rituals of Verification*. New York: Oxford University Press.

Power, M. 2000. The audit society: Second thoughts. *International Journal of Auditing* 4: 111–119.

Power, M. 2003. Evaluating the audit explosion. *Law & Policy* 25(3): 185–202.

Preston, J. & Hammond, C. 2003. Practitioner views on the wider benefits of further education. *Journal of Further & Higher Education* 27(2): 211–222.

Propper, C. & Wilson, D. 2003. The use and usefulness of performance measures in the public sector. *Oxford Review of Economic Policy* 19(2): 250–267.

Rethemeyer, R. & Hatmaker, D. 2007. "Network Management Reconsidered" An inquiry into management of network structures in public sector service provision. *Journal of Public Administration Research and Theory* 18: 617–646.

Robinowitz, D. & Dudley, R. 2006. Public reporting of provider performance: Can its impact be made greater? *Annu. Rev. Public Health* 27: 517–536.

Romzek, B. 2000. Dynamics of public sector accountability in an era of reform. *International Review of Administrative Sciences* 66: 21–44.

Saunders, M., Lewis, P. & Thornhill, A. 2009. *Research Methods for Business Students*, 5th ed. Pearson Education Limited.

Silvia, C. 2011. Collaborative governance concepts for successful network leadership. *State and Local Government Review* 43(1): 66–71.

Skinner, D., Sauders, M. & Beresford, R. 2004. Towards a shared understanding of skill shortages: Differing perceptions of training and development needs. *Education and Training* 4(46): 182–193.

Sobh, R. & Perry, C. 2006. Research design and data analysis in realism research. *European Journal of Marketing* 40(11): 1114–1209.

Stoker, G. 1998. Governance as theory: Five propositions. *International Social Science Journal* 155: 17–28.

United Nations Research Institute for Social Development (UNRISD) 1999. *The New Public Management Approach and Crisis States*. Discussion Paper No. 112, September 1999.

Werner, R. & Asch, D. 2005. The unintended consequences of publicly reporting quality information. *Journal of American Medical Association* 10(293): 1239–1244.

Crisis management and business continuity in the Kuwaiti oil sector

Mohammed Al-Tahous
Ahlia University, Kingdom of Bahrain

ABSTRACT: The oil and gas sectors have performed tirelessly to prevent risks and execute business continuity and crisis management plans within global standards. As the global population continues to rise, the worldwide appetite for energy develops; hence the oil and gas sectors must strive for unique and creative ways to tap into the natural resources of the earth. Business continuity and crisis management are considered to be two of the most reliable strategies that help oil and gas companies survive disasters. This paper discusses the need for crisis and business continuity management during emergencies, contingencies and disastrous conditions with respect to the oil and gas sector of Kuwait. In order to attain the research objectives, a qualitative approach will be adapted to collect primary data from the managers of ten reputed oil companies in Kuwait through interviews.

1 INTRODUCTION

1.1 Research background

The petroleum industry in Kuwait is the biggest industry in the nation, representing around half of the gross domestic product. Kuwait has proven reserves of crude oil of 104 billion barrels comprising 10% of world's reserves. The oil reserves of Kuwait are the fourth biggest in the world and the country is the world's eleventh-biggest producer of oil. The production of oil in Kuwait accounts for 7% of worldwide oil production. Oil has been and still continues to be Kuwait-state's most precious natural resource and the country has the production of oil as its economy backbone, and that even a temporary disaster in production can have severe issues for the economy.

Business continuity management (BCM) is an inevitable entity in Kuwait's oil and gas sector. Business continuity management is a domain of management that concentrates on the development of plans, capabilities and strategies that offer security or alternative operation modes for those business processes or activities which if they were to be broken would bring about a severe business or essentially fatal loss to the oil industry. A developing number of oil and gas companies and their executive management are identifying the significance of business continuity management as part of their enterprise risk management. Crisis management is an essential business continuity management element. A well-structured business continuity management can hinder disasters which consist of financial losses and damages. Accordingly, the crisis management and business continuity management disciplines are succeeding progressively in Kuwait. Business continuity management recognizes critical processes of business and availability hazards, challenging those critical methods. Based on the assessment of risk control, measures are set up to hinder such disasters and to lower damages emerging from such disasters. For this purpose the oil sector in Kuwait has business continuity and crisis management plans in place, which supported them in preventing the destruction and resuming the operation of business within the acceptable down-time period.

The major pre-requisites for a successful business continuity programme involve three factors: first, the determination of those operations and processes critical to accomplish the goals of business and manage essential relationships of customers; second, a core understanding of business surroundings; and third, recognition of critical path operation, inter-dependencies and bottlenecks of process. Thus it can be inferred that the effective business continuity and contingency planning must be implemented by a capable, robust and strong team who can support their oil and gas industry and assure them if any disaster occurs that they have appropriate staffing specialists in place to handle such disasters.

1.2 Problem identified

The oil and gas industry applies several measures to avoid major asset disasters and failures. Leaks or explosions in pipelines, other gas leaks, plant explosions and so on do, however, occur around the globe. The disasters have huge implications for oil and gas industries involving major financial loss and deteriorating reputation. The speed with which modern business is changed shows that a disaster of only a few hours can have a catastrophic influence on the reputation and profitability of an affected organization. Although this will have an adverse and immediate influence, it can also harm the oil and gas industry's big term viability.

Business continuity management will solve these disaster issues and will help businesses to return to their original positions. Business continuity management is a process of holistic management that recognizes essential threats in the oil sector and offers a structure for enhancing resilience of the sector with the capability of an efficient response that protects the concerns of major stakeholders, brand, value-creating activities and reputations of organizations (Opscentre 2003). A developing number of oil sectors are identifying business continuity management as a mission-critical function. Business continuity management is a major concern for oil and gas sectors and it is essential to handle financial reputation and confidence of the business. The oil and gas industry must have extensive business continuity and crisis management as well as an efficient loss prevention strategy to assure uninterrupted production with a prompt recovery and remediation. This paper discusses the significance of business continuity and crisis management plans in the oil sector in case of a disastrous event and provides appropriate measures to avoid such disasters and lead a successful oil sector.

1.3 Aim of the study

To investigate the challenges encountered when implementing crisis and business continuity management during emergency and disastrous conditions with respect to the oil sector in Kuwait.

1.4 Research questions

The following are questions to which this research intends to find the answers:

i. What are the environmental, political and organizational conditions that influence the oil sector in Kuwait?
ii. What are the benefits of business continuity and crisis management with respect to the oil sector?
iii. What are the challenges encountered by the oil sector in implementing business continuity and crisis management in a Kuwaiti context?

The following section gives a review of the relevant literature on business continuity and crisis management in oil and gas industries.

2 LITERATURE REVIEW

Crisis management gives the pre-established guidelines and activities of an organization for preparing for and responding to catastrophic incidents or events in an efficient and safe way. A successful plan for crisis management involve programmes of organization such as disaster recovery, business continuity, communications, emergency response and risk management among others. Crisis management is very important in developing the capability of an organization to react flexibly and is useful in making essential and prompt decisions when a crisis occurs. If an organization is ready for worst-case scenarios then it can manage other circumstances as well. Several studies have been carried out to examine the crisis management role in highly risky industries. Specifically, oil and gas industries have been researched by several authors since crisis management is very important for the oil and gas industry to avoid disastrous incidents. Some off the cuff studies of crisis management in oil and gas industries have been discussed below.

An expert system for decision making assures safety in oil and gas companies. Then the participation of large people groups in crisis circumstances is the key for success in crisis management leading to an appropriate integration of planning, critical analysis, actions based on knowledge, flexible adaptation and group communication. Supporting these characteristics in mind the AI (artificial intelligence) community regards scenarios including these crisis situations as an ideal verified for their techniques. Oil and gas industries all over the globe are greater risk industries due to the critical conditions of working and their nature. The oil and gas industry is one of the industries that has greater risk factors and has greater injuries and fatalities in the workplace. Sawalha (2011) examined the significance of BCM in the terms of SP (strategic planning) that is combining BCM with SP in a single framework. They collected sample data from four sectors namely services, banking, industrial and insurance. The findings of this study indicate that 81% of the Jordan organizations use BCM and 52% of the organizations use BCM with respect to strategic planning. Strategic planning was considered as the most significant driver to achieve the organizational purposes that are associated with BCM. Business continuity management must be viewed as a greater level function of management that has the importance to play an essential role in accomplishing success in organizations. Organizations that acquire business continuity management in a strategic view can recover from crises swiftly with small influence on their rivalry position and good practice proposes companies to place business continuity management at the middle of organizations' strategic and cultural objectives. Thus it is understood that the BCM programme benefits the organization to minimize risk, maintain and protect customers, ensures long-term organizational survival, prepares for unexpected crisis and disasters, understands the environment in the business and helps in developing planning processes and protecting assets. Similarly Mukerjee & Gupta (2010) developed a modelling framework for estimating smart-infrastructure crisis management systems' effectiveness. This analysis developed the (CRM) criticality response-modelling framework application on fire emergencies on OGPP (oil and gas production platforms) and stimulation based verification by estimating the process of emergency response using CRM through OGPP. It is inferred that the framework of CRM is used to analyse the comparative study on the result of various response actions and from this study the criticality response-modelling framework provides a smart infrastructure which gives good crisis preparedness. Similarly, offshore operators of oil and gas projects of BCM focus on implementing, testing and developing strategies and plans that will assure that, in the event of an incident affecting information of technology and affecting loss of office premises, an efficient recovery will be handled and crucial activities of business will be managed.

Contrary to that Musgrave & Woodman (2013) conducted a survey to examine the effectiveness of the BCM. In this study business continuity management is helpful to lower disruption, speed recovery and provides benefits which assure the costs. This article helped to create a strong case of business for business continuity management as a way of supporting managers satisfying their liability to keep their businesses functioning. Thus it was concluded that every organization must have a robust and distinct proportionate approach of business continuity plan for the oil and gas sector. Consequently Heller (2012) in his article describes the problem of crisis management of organization within British Petroleum when the deepwater horizon of British Petroleum oil rig break in the Gulf of Mexico with an explosion that killed eleven rig workers and triggered North America's biggest environmental disaster. Fink (1986) has mentioned that BP is now in a varied position in the stage of chronic crisis due to their management failure to enhance a timely response to vehicles' malfunctions. This study was a good example for mismanagement of crisis by a familiar leader in the oil and gas industry. It was understood from the study that the crisis management plan is very helpful to take actions and decisions against crisis to restore the global reputation of British Petroleum.

According to an on-line Technical Information Paper (2013) an efficient response to an oil spill relies to a large extent on the preparedness of individuals and organizations involved. Responding to a severe spill of oil affecting a vast number of organizations and people demands that different decisions are made rapidly. This can be accomplished only if all respondents are prepared adequately to make difficult decisions when needed and can mobilize proper resources with reduced delay and hesitation. A completely developed contingency plan is used in organizations to accomplish targets.

Insurance is useful and accessible for oil and gas companies which face spills but it does not secure them from the issues of public relations that exist. Contrary to that, the Exxon Valdez oil disaster which led to large scale pollution and lost wildlife affected by the oil spill as an example of outcomes that cannot be changed by any insurance payment level. In certain cases executives of businesses trust that insurance is the solution for the issues of crisis management. It can be inferred that crisis management is very important for oil and gas industry to solve the crisis issues when oil spills happen but insurance was a safeguard to salvage and prevail the oil and gas industry business.

Despite each endeavour to hinder them, the spills of oil still happen.

It can be concluded that business continuity and crisis management planning is very important to respond efficiently to emergency situations and to avoid an essentially serious disturbance in their oil and gas operations.

3 RESEARCH METHODOLOGY

3.1 *Research paradigm*

According to Cooper & Schindler (2010), research paradigm is the first step in any research. Paradigm is the key component of any research and the process will be continued after the paradigm formulation. Paradigm is called 'the lane of perceptions and values'. Paradigm is of two types: positivism and interpretivism. Positivism is the common approach and it is used to collect data from a large number of samples. Interpretivism is the unique technique that is followed to particulate the problem and the solution from the strong and close views.

This study uses the interpretivism technique. Crisis management in the oil sector is a very complex problem for those who are engaged in it. Therefore, the study will interview the managers who are in the crisis management process.

3.2 *Research approach*

Research approach is the very next step in the research investigation. It is the proceeding of the research after the paradigm is fixed. It is divided into qualitative approach and quantitative approach. Interpretivism is the qualitative approach and the quantitative approach will be positivism (Yin 2009).

This study follows the qualitative approach. The interview technique will ensure the quality of the data and the data collection will be through a discussion or communication between the researcher and the respondent. This qualitative approach is about the study of natural things and the investigator will not change the actual settings of the study.

3.3 *Research design*

Research design is said to be the guide for the research process. A research method that is full of meaning and examination of a real-life modern occurrence in its usual background is a case study (Yin 2012). Research design is classified into exploratory research design and conclusive research design. Conclusive research design is further divided into descriptive and causal research designs.

This study follows the descriptive research design. According to Best & Kahn (2007), the term descriptive research has often been used mistakenly to describe three types of investigation that are unalike. Possibly their exterior similarities have hidden their distinction. The descriptive research design will describe the problem and hypothesis.

3.4 *Sampling design*

Frey et al. (2000) define that sampling design or the sampling plan is said to be used in the sample selection from the vast number of respondents. The techniques in the sample design will intend to lead to a decision on what basis the respondents are to be selected. Sampling techniques are divided into two major segments. They are probability and non-probability sampling techniques.

The probability sampling technique will comprise the following divisions: simple random sampling, stratified sampling, systematic sampling, cluster sampling and multi-stage sampling. Non-probability sampling is divided into convenient sampling, judgmental sampling, snowball sampling and quota sampling.

This study employs quota sampling as the sampling technique. Quota sampling is the famous non-probability sampling technique that is widely used in research which van Maanen (1983) describes as the one where respondents are selected non-randomly on the basis of their known amount of the population. Quota sampling will select the samples like the stratified selection. It first segregates the population on the basis of some groups and then it will select the respondents from the strata.

The sampling unit for the study is the following oil sector firms in Kuwait

   i. Kuwait Oil Company (KOC),
  ii. Kuwait National Petroleum Company (KNPC),
 iii. Kuwait Foreign Petroleum Exploration Company (KUFPEC),
  iv. Kuwait Petroleum International (KPI),
   v. Kuwait Gulf Oil Company (KGOC),
  vi. Oil Support Services Company (OSSC),
 vii. Kuwait Aviation Fuelling Company (KAFCO),
viii. Kuwait Oil Tanker Company (KOTC),
  ix. Petroleum Industries Company (PIC),
   x. Oil Development Company (ODC).

The target population for the study is the ten key managers or supervisors of the selected firms from the oil sector in Kuwait. The sample size of the study is the ten managers or supervisors of the oil sector in Kuwait. The data collection is planned to gather information regarding the crisis management of the oil sectors through the distribution of open-ended questionnaires. The data will be collected from the ten managers or supervisors of the Kuwait oil-sector firms. Questionnaires will be distributed to the respondents at their places of work. They will ask the following questions:

   i. Has your organization encountered any disasters in the past?
  ii. How did your organization overcome those disasters?
 iii. Why do you think that BCCM is necessary for oil and gas sector?
  iv. What are the BCCM strategies adapted at your organization?
   v. What are the benefits gained by your organization on applying BCCM strategies?
  vi. What are the challenges faced by organization in successful implementation of BCCM strategies?
 vii. Please comment if appropriate.

3.5 *Data collection method*

Research will be accurate and particular only in the compilation of data. The task of data collection starts only after the planning part of the research problem. Data collection can be carried out in two ways: primary and secondary data collection (Smith 1983).

Primary data will give the exact solution for the problem. These data are collected by the researcher's own effort. The primary data will be collected and used for the first time. In this study, the primary data would be collected in the form of interviews using open-ended questionnaires through the internet. Secondary data are the facts which were previously present in some form or another, but which were not collected for the first time. Secondary data are often the starting point of data collection in as much as they are the first type of data to be collected. This study will collect the secondary data from the secondary sources of information such as the web, firms' reports, journals, articles, etc.

3.6 *Analysis and interpretation*

Tashakkori & Teddlie (1998) infer that the analysis and interpretation of data will be the most important part of any research. This section will make the evaluation of collected data. Each research study must be carefully planned and operated according to particular guidelines. When the analysis is finished, the researcher must step back and assume what has been deduced. This study is carried out with the interpretivism technique and it has to be evaluated by the text analysis. Text analysis will be the perfect tool for the qualitative analysis of data.

3.7 *Strategies for validation of techniques*

The qualitative study will follow the validation techniques like credibility, conformability, dependability and transferability. This research makes sufficient place for the mentioned validation techniques.

3.8 *Ethical considerations*

The study follows the basic ethics that have to be inclined to the study. Ethics are the components of a study that reflects the success of the research.

3.9 *Summary*

The research follows the interpretivism technique. The research approach followed is the qualitative approach, the research design is the descriptive research design, and the research has used the quota sampling technique for the respondent's selection. The study is carried out in ten oil sector firms of Kuwait. The respondents are the managers or supervisors of the target firms. The data will be collected through open-ended questionnaires. The study follows the validation techniques and the ethical considerations. The study also mentions its limitations.

4   DISASTERS IN THE OIL SECTOR

4.1 *Types of disaster that are common in the oil sector*

When managers were asked about the types of disasters in oil sector one of the managers from Kuwait Aviation Fuelling Company mentioned that acidification in the ocean as one of the disasters in the oil sector. The resources of ocean with 5% of worldwide population but 25% of its production of greenhouse gases are the major cause of altering oceans' chemistry. These alterations are dissolving the clam shells, lobster, shrimp and affecting other influences to fisheries worldwide that dominate the issues affected by the oil and gas sector. Similarly one of the managers from Kuwait Foreign Petroleum Exploration Company pointed out that leaking tanks and pipes as one of the disaster in oil and gas industry because deposit deposits nearly 252 million gallons of chemicals and oil into aquifers below the refinery. To provide clean water to the oil and gas sector, fossil fuels are burned to produce enough electricity to pump water hundreds of miles across land adding to asthma, state debt, air pollution and global warming.

It can be inferred that the above mentioned disasters all resulted in loss of businesses, sustainable jobs and productivity of workers and they must be prevented from occurring.

4.2 *Disasters encountered by organizations in the past*

One of the managers from the Kuwait Oil Company mentioned that the organization has encountered disaster in the past. In the year 1990 the former regime of Iraq occupied Kuwait and caused devastating and heavy destruction throughout the nation's different industrial, infrastructure and economic sectors. Kuwait Oil Company was the most ruined sector of all and the measure of damages caused at facilities and installations of oil was unprecedented in the evolution of the modern-day oil sector.

It can be inferred every may experience any disasters in the past and so each organizations is now taking essential steps for recovery.

4.3 *Process of overcoming disaster*

It was understood from the study that one of the managers from the Petroleum Industries Company mentioned that their organization overcome the disasters by supplementing the programme of risk management with prescriptive pollution and safety prevention standards to interior department that are developed and chosen in consultation with peers of international regulatory and are careful as the terms of leasing and regulatory needs in peer oil generating countries. Performing with the Forum of International Regulators, the interior department must recognize those production, drilling and standards of emergency response that best secure the offshore employees and surroundings and begin new revisions and standards to fill the space and proper deficiencies. Standards must be updated every five years under the process of formal review of ISO (International Organization for Standardization).

Contrary to that, one of the managers from the Kuwait National Petroleum Company mentioned that it is essential to check often for leaks, decay and deterioration in all equipment. This must be performed on a yearly basis. Their organization has made an independent agency within the interior department with enforcement authority to visit all perspectives of offshore drilling protection as well as operational and structural integrity of offshore production of energy facilities.

It can be inferred that the oil and gas industry must seek to increase extensive overhauls of regulatory and leasing policies, and organizations must oversee activities of offshore plant regularly to avoid disasters.

4.4 *Need for BCCM in oil and gas sector*

One of the managers from the Kuwait Foreign Petroleum Exploration Company said that the oil and gas industry is facing the challenge and requirement to develop business continuity management plans further. Ranging from cyber security attacks, control of wells, natural disasters and other issues, the oil and gas sector has viewed a development to refurbish all efforts of emergency management. Oil and gas sectors are to enhance and further develop existing methods that will aid in handling risk seamlessly. Due to the impact of such incidents building business continuity management plans is very essential to restore the operation and maintain personal safety. One of the managers from the Oil Development Company described that business continuity management as the method through which organizations target to continue to provision of their major services and products during and following a disruption to normal activity and to influence a recovery afterwards. He also added that business continuity management will help to hinder crises, disasters and operational interruptions and will support organizations to return rapidly to a state of business as usual if any of the incidents exist.

It can be inferred that effective business continuity management is the first defence line for oil and gas industry to maintain the provision of their core services and to assure their operations' survival.

4.5 Organizations' BCCM strategies

One of the managers from Kuwait Oil Company mentioned that their organization adapts business continuity preparedness. The crisis management and business continuity management disciplines are developing in Kuwait Oil Company progressively. Though business continuity management preparedness and capabilities are set up within an organization they perform in isolation and at times are restricted to technological recovery. Similarly, one of the managers from Oil Support Services Company said that their organization's BCM perceives how loss of resources could influence capability and to recognize and execute proper risk contingencies and controls pre-event. Their organization protects their operations, shareholder and credibility value using business continuity management.

It can be inferred that BCCM services help in examining, implementing and developing extensive plans and strategies of business continuity to assure accessibility of difficult processes of business.

4.6 Benefits rendered by BCCM strategies to an organization

When the managers were asked about the benefits gained by their organizations using BCCM, one of the managers from Oil Support Services Company pointed out that business continuity management will strengthen the organization not only against big scale issues, it can also help in little issues that might have affected interruptions of continuity to become moot through brief planning. Similarly one of the managers from Kuwait National Petroleum Company pointed out that business continuity plan secures the brand, reputation and image of an organization and it also reduces the losses following disasters. Contrary to that, one of the managers from Oil Support Services Company has mentioned that business continuity plan responds effectively and rapidly to an adverse circumstance so that oil and gas sectors can move one step ahead of their rivals when it comes to satisfy the needs of tender. Business continuity management also lowers insurance premiums, which is one of the reduced known advantages of BCCM.

It can be inferred that bringing business continuity management into practice in oil and gas sector can make the business for essential disaster which assures that they will be able to manage continuity of their business practices and remove or reduce the influence disasters have on the oil and gas sector.

4.7 Challenges faced by organizations in the successful implementation of BCCM strategies

The last question asked to managers was about the challenges faced by organizations in using BCCM. One of the managers from Kuwait Aviation Fuelling Company mentioned restraints of budget as the challenge in using BCCM because oil and gas sector makes gains by mitigating and planning measures which are always compromised from other priorities. It may be helpful to evaluate the implementation cost for every difficult process in relation to price of a difficult process reduction. This exercise may indicate the requirement for a designated budget. It may be essential to prioritize implementation of business continuity management by every difficult process with a step-by-step timeline for completion. Similarly, one of the managers from Kuwait Oil Tanker Company pointed out preparedness culture as one of the challenge in using BCCM. Employees who are skilled in procedures of business continuity management will be ready in the incident of an operational failure. Managers who embrace and emphasize continuity and safety measures will make a work environment that reflects those measures and manages an overall preparedness culture.

It can be inferred, therefore, that the oil and gas industry depends on under-tested or untested continuity-related methods to handle the challenges of business interruptions.

5 SUMMARY

It can be concluded that the environmental circumstances which are disastrous for the oil and gas industry are oil spills, emissions to air and emissions to sea. The major disaster which can exist in oil and gas industry will be oil spills. Most of the disastrous spills are little oil spill. The most frequent effects of disaster are failure in equipment, mistakes committed by employees and extreme conditions of the environment. The environmental circumstances of disastrous discharges are severe when they exist near to shore or in remote places. Blowouts in the oil and gas industry can cause huge damage to drilling rigs to and to drilling personnel on rig as well as environmental damage. Similarly, pollutants emission into the surroundings is the major disaster in oil fields. The most vast spread source of such emissions is huge amount of hydrocarbons in testing course and exploiting wells and burning of casing head gas. Venting and flaring of natural gas in wells of oil is an essential source of emissions of greenhouse gas. The emissions-to-sea disaster points to existence of oil and gas structures that produce solid and liquid waste. This waste comprises geological material in the form of spent drilling fluids and mud coming out from well. Drilling fluids usage with a base of petrol is very dangerous to the environment which may damage the fishing industry in that area. Thus government must provide huge funds for oil industry stakeholders to solve the environmental disasters in oil and gas sector.

It can be concluded that business continuity management system is needed essentially for oil and gas industry on how to handle reviews, activities and exercises of training so that the major processes of business are secured and defined against threat in case of disasters. The most immediate and greatest value of the BCCM process is the awareness which an individual acquires of the information of her or his business and not essentially the streamlining of how to manage disaster as an organization. Business continuity planning always creates awareness of helpful steps to develop oil and gas industry which had not been considered previously. Business continuity planning is much smarter than oil and gas industry business because it helps the organization remain positioned better to recover from the interruption of business, financial influence, damage of assets and life loss which a natural or manmade disaster may cause. Thus having a well documented and pre-defined business continuity and crisis management plan is necessary for the oil and gas sector to interact on how their companies will respond during a disaster and is one of the best investments which the oil and gas sector can make to make their company profitable and successful.

It can be concluded further that the benefits of business continuity and crisis management are security of assets through reduction of risk, better avoidance of liability practices, security of markets by helping to assure provision and reputation security and compliance with safety and health legislation. The oil and gas sector acquires the decisions of employees on the actions taken by organizations and assures business continuity if disaster strikes. Business continuity management is very helpful in emergency issues of organization and maintains services delivery in case of incidents. The business continuity management lowers premiums and access to finance better which give a commercial benefit to oil and gas industry before a disaster happens. Business continuity management enhances the confidence of workers in the fact that the business is performing well. It drives efficiencies of operation and adds competitive benefit. Quick response to a crisis or disaster develops business reputation and develops customer confidence. Thus every oil and gas industry must approach business continuity management and enable effective and rapid crisis management in one solution.

The challenges encountered by the oil sector in implementing business continuity and crisis management are lack of management assistance, lack of business continuity and training awareness, turnover of employees, accomplishing a stable readiness state, recognizing critical processes and protecting suppliers for business continuity. It is challenging to carry out a cost benefit examination for continuity of business. The decisions of management are usually based on existing financial aspects that advantage stockholders, bottom line and departments. Offering corporate decision makers and managers a brief hazard-and-vulnerability examination with visible statistics of finance of their impacts may provide some assistance. Training must provide procedural flexibility based on disaster demands, continuing assessment and offer choices for every scenario. Similarly, employees must have distinct knowledge or experience of business continuity that can be used to develop the plan of business continuity. The processes of business continuity can be executed as part of SOP (standard operating procedures). A stable continuity level can be sustainable if a personnel, process or facility is not accessible. The capability to quantify and examine which critical processes of business when not functionally may harm the reputation of organization or ability to perform is a difficult stage in the process of business continuity management. The oil and gas sector must examine alternative supply arrangements that can directly reduce the disruption threats. Thus a well-developed plan of business continuity management can reduce disasters in oil and gas industries while protecting major interests, assets and relationships of business.

REFERENCES

Best, J.W. & Kahn, J.V. 2007. *Research in Education*. New Delhi: Prentice Hall of India.
Cooper & Schindler 2010. *Business Research Methods*. New York: McGraw-Hill.
Fink, S. 1986. *Corporate Crisis Leaders: Planning for the Inevitable*. New York: American Leaders Association.
Frey, L.R., Botan, C.H. & Kreps, G.L. 2000. *Investigating Communication: An Introduction to Research Methods*. Boston, MA: Allyn and Bacon (second edition).
Heller, N.A. 2012. Leadership in crisis: An exploration of the British Petroleum case. *International Journal of Business and Social Science* 3(18).
Lockwood 2005. Crisis management in today's business environment: HR's strategic role. *SHRM Research Quarterly*.
Mukerjee, T. & Gupta. K.S. 2010. A modelling framework for evaluating effectiveness of smart-infrastructure crises management systems. Retrieved on 18 February 2014 from http://citeseerx.ist.psu.edu/viewdoc/download?doi=10.1.1.139.5216&rep=rep1&type=pdf
Musgrave, B. & Woodman, P. 2013. *Weathering the Storm: The 2013 Business Continuity Management Survey*. London: Chartered Management Institute.
Opscentre 2003. Business continuity and crisis management. Available at http://opscentre.com/resources/pdfs/Research/Business%20Continuity%20and%20Crisis%20Management.pdf [accessed on 05 March 2014].
Sawalha, S. 2011. Business continuity management and strategic planning: The case of Jordan. Huddersfield: University of Huddersfield.
Smith, J.K. 1983. Quantitative and qualitative research: An attempt to classify the issue. *Educational Research* March: 6–13.
Tashakkori, A. & Teddlie, C. 1998. Mixed methodology: Combining qualitative and quantitative approaches. *Applied Social Research Methods* 46.
Technical Information Paper 2013. Contingency planning for marine oil spills. Available at http://www.itopf.com/information-services/publications/documents/TIP16ContingencyPlanningforMarineOilSpills.pdf [accessed on 05 March 2014].
Van Aanen, J. 1983. *Qualitative Methodology*. London: Sage.
Yin, R.K. 2009. *Case Study Research: Design and Methods*. Thousand Oaks, CA: Sage Publications (fourth edition).
Yin, R.K. 2012. *Applications of Case Study Research*. Thousand Oaks, CA: Sage Publications (third edition).

QUESTIONNAIRE

Crisis Management and Business Continuity as a priority involving Emergency, Contingencies and Disastrous conditions at Oil Sector, Kuwait

Personal Profile:

Q1. Name   :
Q2. Age   :
Q3. Education   :
Q4. Work Experience in Years   :
Q5. Designation   :

Crisis Management and Business Continuity as a priority involving Emergency, Contingencies and Disastrous conditions at Oil Sector:

1. What are the types of disasters that are common in oil sector?
_____
_____
_____

2. Has your organization encountered any disasters in the past?
_____
_____
_____

3. How did your organization overcome those disasters?
_____
_____

4. Why do you thinks that BCCM is necessary for oil and gas sector?
_____
_____

5. What are the BCCM strategies adapted at your organization?
_____
_____

6. What are the benefits gained by your organization on applying BCCM strategies?
_____
_____

7. What are the challenges faced by organization in successful implementation of BCCM strategies?
_____
_____

8. Comments if any
_____
_____

Thank you for your time

Challenges faced by female entrepreneurs in developing countries

Najma G.R. Taqi
Ahlia University, Kingdom of Bahrain

ABSTRACT: The findings and conclusions of this paper, which aims to study and understand female entrepreneurs in developing countries, will be of great help in a number of ways. States all over the world are making policies for making it easy to set up business. Such changes could give an advantage to female entrepreneurs considerably. States which rank highest on the easiness of doing business, are related to higher proportions of women entrepreneurs than men. This paper focusses on knowing the reasons for fewer female entrepreneurs in developing countries as compared to male entrepreneurs in these countries. The research shows that females in lower and middle income states are less likely to take part in entrepreneurship for many reasons. So, policies require to be structured for the building of entrepreneurship educational programmes for younger females that encourage them to start entrepreneurial activities as their future profession.

1 INTRODUCTION

Female entrepreneurs of today are so successful that it is simple to assume that high-status business female in various fields have always been a feature of the economy. From a historical perspective, we can see how female business owners have always fought the exclusive challenges given by socially determined sex roles that have both created opportunities for the development of females and restricted their expansion (Storey 2009). Female entrepreneurship is a way out of economic disparities and helps women to be authorities in their own right. Their entrepreneurship contributes in a positive way in different dimensions and facets in economic growth and the creation of jobs (Storey 2009).

Positive projection of activities of entrepreneurs in a state must be a vital aspect of any government agency for boosting economic success. Entrepreneurial behaviour of the female, in contrast to the male, is more likely to be shaped by structure of families and social ties in both higher as well as lower income states. Throughout the world, female generations from diverse backgrounds demonstrate encouraging symbols of entrepreneur strength. It is expected that governments at every level will work towards providing an environment where this strength can grow (Venkataraman 2010). The factors that influence the activities of entrepreneurs are relatively diversified amongst nations, as can be seen by examining their industrial sectors, utilization of technology, firm service, and development and potential.

At present female entrepreneurs represent a women's group who has escaped from the beaten track and are discovering new opportunities of economic contribution. Amongst these reasons for females to manage and start systematic enterprises are their abilities and understanding, their endowments and skills in business and a persuasive aspiration of desiring to do something positive. What makes women's entrance and successes especially important and admirable are the efforts they have to make, aggravation they have to face, and the various hindrances they have to prevail over for emerging as entrepreneurs at the earlier phase, and then achieving achievement in businesses at the point of running their enterprises. Standing of women's independence relies on economic situation even more than political; if the women are not independent and self-earning, they will have to rely on their husbands or someone else, and dependents are not at all free.

"Women in enterprises" is supposed to be a current trend in developing countries. The fact that about half the population of developing countries is women, whereas businesses run and managed by them comprise not more than 5% is a sign of social, cultural and economic interruptions in the years of development. Certainly, the involvement of women in economic activities and productivity of goods and services is far higher than formal figures may show, as much of it occurs in the informal sector as also in the households. Over past years, while females have stepped forward to start their own businesses, their numbers remain fewer and far between. The entrepreneurial world is still under the control of men.

So far, society has persisted in promoting a different misconception (myth): the understanding and doubts about female entrepreneurs in business and industry. This is because of the lack of literature on women in business and industry in developing countries. The lack of evidence, information and documentation regarding female entrepreneurs has permitted a stereotype woman's image in business to continue. Females have been successful in breaking the imprisonment in the restrictions of their houses by taking part in different types of professions and services, and female entrepreneurs have been supposed to be equality to their men counterparts in a business sense and expertise and are coming out as sharp and dynamic entrepreneurs (Carter 2001). There are several reasons for females to take part in entrepreneurial businesses in a predominately society. Entrepreneurship amongst women is a vital opportunity by which they can prevail over their subordination in their families and society as a whole. Thus, entrepreneurship development amongst females has had particular consideration by policy makers. Several organizations, institutions and federations promote and build up female entrepreneurship by giving financial support at concessional interest rates and also systematize business fairs and exhibitions.

2 THEORETICAL BACKGROUND

Large gender gaps in start-up activities are found in developing countries, and they are likely to be narrow in low-income countries possibly as numerous women start businesses out of necessity. Unexpectedly, women in developing states are likely to be more self-confident about their capabilities (expertise and knowledge) for becoming entrepreneurs and not so afraid of failure as compared to women in middle and high income states, despite subjective and probably biased perspectives regarding self-confidence, fear of failure, and existence of opportunities or important and, in a systematic manner, related determinants of the gender gap across every country.

Women in developing states, like their counterparts in more advanced ones, depend more, as compared to men, upon extended families that, in numerous rural settings, are frequently their single or key social network. This is frequently restraining as marriage status of women, and the resources and incomes brought to their marriages, come out as key determinants of decisions of their entrepreneurship. Young females with young children tend to start more as entrepreneurs as compared to waged labour, and are more inclined to be entrepreneurs as compared to unmarried women, though they are also more likely to voluntarily quit businesses (Brush 2010).

As far as the firm's performance of women entrepreneurs is concerned, the facts from developing and developed states are to some extent the same. Females are likely to have low development expectations and their businesses are likely to develop slowly in both sales as well as employment as compared to those of males. Evidence suggests that the main concern of women in various developing states is not with development but to a certain extent with continued existence (Carter 2001). This can be a reason for the finding that usually women entrepreneurs in developing states is likely being portfolio instead of sequential entrepreneurs, as they try to expand income sources and continued existence possibilities.

A woman faces a number of challenges such as illiteracy, poorer accessibility to education and recognition, gender-based aggression and inbuilt discrimination that limit both their economic as well as political opportunities. Women also tackle corruption, biased rules and legislations, and lack of heritage and property rights (Myers 2005). Once they are given the opportunity to take charge of their economic as well as financial securities, women not just take their destiny in their hands but also transform the trajectories for their families as well as their communities.

Female entrepreneurs are potentially stimulated to change so as to improve living conditions for their families, give a sharing hand to their husbands in producing income, give quality education to their children, contribute optimistically by creating job opportunities, giving power to other females. Also, the literature has discussed the female entrepreneurs' fear of failure in lower or middle income states and is supposed to be high because of the current conditions amongst females in these states. Females in the European and Asian lower/middle-income states have higher rates of fear of failure (40.3%), than that of Latin American and Caribbean lower or middle income states (34.2%) and females in higher-income states (27.1%) (Venkataraman 2010).

Women entrepreneurship has become well-known all over the world in different contexts. Amongst them, three vital contexts are easily recognized: efficiency increase in market places, poverty alleviation as well as social empowerment of women. Such contexts describe practice areas and various aspect concerns may be looked through. Likewise, such contexts signify diverse priorities in the field of support policies of women entrepreneurship.

Three functions of women entrepreneurship stated by different actors in giving reasons for their activities in the area are: increase of participation of women in market place production, thus solidifying the area of gender equal opportunity, reducing poverty and participation in state growth. The foremost context that woman entrepreneurship issue has been placed in is the policies of the labour market. Support policies of woman entrepreneurship have increased in popularity as the way to develop employment of women. The long-term purpose of supporting women within the labour market has found itself a strong tool set with active labour market policies (ALMP). These policies were programmes for intensive employment searching, development training of human resources, short-term work plans and business venture growth. Business venture growth support was also the most accessible approach amongst others for a few reasons.

Thorough discussions have occurred on motivational factors that affect females to join the important group of entrepreneurs; they are of two kinds. First is entrepreneurship by choice, and the second is entrepreneurship by need. A woman becomes an entrepreneur by choice because of the following factors:

- for materializing their ideas into capital,
- for their authorization and freedom,
- for proving their value amongst their male family members,
- for establishing their own regulations for their work,
- for overcoming the insufficiencies they experienced in their job experience,
- a durable standing wish to own their own firm,
- working for somebody else did not attract them,
- by need they are motivated to be entrepreneurs are,
- improving the quality of life of their children,
- sharing the economic burden of the family,
- adjusting and managing family and business life in a successful way on their own terms,
- the death or illness of their husbands.

Women are not always observed as agents for growth in development programmes although there is a noticeable change in gender regulations and gender growth aims of governments. In remote areas in developing countries, most bank officials are unwilling to provide loans to females in distant districts and several have not heard of different schemes of finance introduced for female entrepreneurs. In these regions, husbands reject to be guarantors for their wives who seek loans.

Investing for females is a higher-yield investment. Gender equality in accessibility toward education, health care, political involvement and economic contribution is the key to competitiveness and success of a country. Women may create twofold dividends.

Women entrepreneurs are more likely to employ other women. Additionally, to create employment opportunities for women, such a trend will lead to a general increase in the development of women in developing countries. Better contribution of women in remunerative works will not just improve their living circumstances but also their bargaining status. Employed women will have good accessibility for mainstreaming banking services that will assist them saving and investing their incomes in tangible resources.

There are various obstacles a female may face when being involved in a process of entrepreneur. This part of the discussion may be divided into four parts:

- general barriers to a female involving in process of entrepreneurs (opportunity identification and inclination to establish businesses),
- particular barriers to create (collecting essential information, financially as well as human resources for starting a business),
- particular barriers to manage a small business,
- and particular barriers to growing businesses.

General barriers include lack of connectivity, lack of experiences, lack of appropriate networks and societal position, lack of wealth, vying demands in time. Particular barriers include barriers particular to starting a new firm, barriers particular to managing a small firm, and the particular case of the family businesses.

We can see a very important discussion about female entrepreneurship in the literature section as it demonstrated different psycho-social and socio-cultural factors (Carter 2001). These factors serve as obstacles to the entry of a female into entrepreneurship. A female has restricted ideas of business, opportunity, and is not conscious of sources of support help. A female frequently begins with lower confidence, lower self-assurance, and because of this, she lives her life limited to the four walls of her house (Myers 2005). It is against this context that a female entrepreneur enters the business world, lives, operates or exists. In circumstances like this, support of the family is of huge importance in the process of decision making for female entrepreneurs and relying on reaction from families, female entrepreneurs can either be extremely motivated in their business or totally de-motivated.

The need of today in developing countries is to encourage development in such a manner that the most important preference is given to skill growth as well as the education of women. Women have been forced into the field of small business venture as entrepreneurs. Throughout the past two decades, the developing world has been successful to a greater extent in promoting the development of smaller businesses, by a package of support initiatives at different levels (Carter 2001). They consist of policy, facility services, finance, infrastructure as well as training. The promotional agencies are decisively determined to turn the smouldering fire into flames. Hence, with the promotional agencies' dynamic support as well as better family professional background, women entrepreneurship amongst the comparatively well educated women is fast developing and it is likely to build up in every part of the countries in years to come. Women entrepreneurs are being assisted in recognizing and adopting recently developed projects. Term loans are given by financial organizations on a liberal basis. A new package of dispensation and incentives, as well as financial support for the promotion and growth of women entrepreneurs has been initiated. Special incentives are being given to women entrepreneurs by several countries priorities is given just for those women entrepreneurs during selection for training of entrepreneurship who have a good background of education, special skills for entrepreneurial training, business as well as craft enterprise experience, trade and business, background of their families, and so on.

3 STATEMENT OF THE PROBLEM

Entrepreneurship is one of the vital factors of industrialization; without entrepreneurship, industrialization is impossible. Entrepreneurs play a significant role in the economic development of developing countries. Furthermore, entrepreneurship has played a vital role in developing societies of fast developing states (Brush 2010). Currently, developing economies have been understood that enterprising females have entrepreneurial abilities that could be harnessed in order to change them from the status of "job seekers" to "job givers". Governments have realized the significance of female entrepreneurship and consequently offer many programmes for female entrepreneurs. In the developing world, many entrepreneurs do extremely well in small scale industries. Although the governments organize females by different associations, they are not willing to start the business. Women are less motivated to establish business units than men because of unnecessary fear and lack of motivation.

It was just in the last few years that more scientific-based studies of female entrepreneurs were carried out in different geographical areas in developing countries that say about motivations and potential barriers to women's participation. According to these studies, at present, women are facing barriers like gaining finances, work-home disputes, need of education as well as training in business and management skills. They showed, moreover, that financial assistance, business training and guidance, the need to network with different business owners as well as marketing support, are their key support requirements. There is sufficient justification, against the unproductive background of scientific value-added research, for pursuing deeper investigations into the challenges faced by female entrepreneurs in developing countries.

4 SIGNIFICANCE OF THE STUDY

The Ph.D. study associated with this paper will help to realize different barriers, issues and challenges faced by women entrepreneurs and it will also contribute to make realized policy makers to take steps to solve these problems and to motivate them by different measures. It will make a contribution to the database of female entrepreneurs (Heemskerk 2011). The study explores the context of female entrepreneurship with particular focus on developing countries. This study can give help to government officials and policy makers as well as other governmental and non-governmental agencies that function for female entrepreneurship growth. The research can be helpful to female entrepreneurs themselves to develop their businesses into successful ventures.

5 FURTHER RESEARCH

This section of the paper outlines some areas in which further research is necessary.

Firstly, we require more theory, as theoretical development has not been equal with the huge sum of empirical studies.

Secondly, a considerable and so far unsettled issue concerns what variables must enter the effectiveness function of individuals when examining their distribution of time between household production, paid labour and self-employment, mainly in developing countries, and when alternative perspectives of the family unit are taken into account. When implemented to sequential entrepreneurship, the theoretical and empirical literature has very little to say on females in developing states.

The third point is that questions about cultural factors and relocations amongst the self-employed give a different extremely productive area of investigation for both theory and empirical work, with the opportunity to make not just a considerable contribution to science but also to policies and management practice.

Fourthly, discrimination has been identified as a possible reason for the gender gap in entrepreneurship; this is more obvious in developing countries, though the facts are mixed. Discrimination against females is frequently the result of gender thinking intrinsic in the cultures or societies (Carter 2001). This can have the impact of not just minimizing the chance of women of being entrepreneurs with equivalent earnings as, but can also decrease the non-pecuniary advantages females get from entrepreneurship.

The fifth point is that extremely little is known about how the level of aggregate activities affects the decisions of women regarding entrepreneurship and not much is known regarding how the latter contribute to development. Though a considerable sum of subjective evidence and a few extremely good case studies are present on this topic, the lack of a methodical approach and data has avoided, to date, the formulation of an inclusive and strong theory of women entrepreneurship and development. Certainly, no 'female only' theory is required but a strong understanding of how the unique attributes of women entrepreneurship are considered by existing models of development would be highly attractive for both science as well as policies.

Lastly, the study of institutes and how they encourage or discourage women entrepreneurship is mainly required for its policy implications, particularly in developing countries in which issues of development of institutions has in current years been emphasized. In this context, a post-institutional approach on the basis of insights from economics and organizational theory appears favourable and economic approaches which incorporate tools and techniques from anthropology (the study of humankind) and ethnography (the scientific explanation of peoples and cultures with their customs, behaviour and common differences) could form interesting studies.

6 CONCLUSIONS

Looking further on, while accessibility to labour markets is growing for women, capacities developing services for women workers like inclusive training programmes require further development in developing countries. One of the key challenges faced by most female entrepreneurs is the lack of accessibility to marketplace information. So, supportive infrastructure, like information centres, have to be founded for disseminating information on potential buyers, existing technologies and finance that will make their businesses sustainable & profitable.

For females being empowered, they require having equal accessibility to education and healthcare, and the chance to start businesses. Women invest 80% of their income in their families and communities, and women-led local governments are both more truthful and give better public services. When governments invest in education of women and give them the opportunities to have accessibility to credit or start the smaller businesses, the economic, political and social advantages ripple out far away from the house.

What common sense tells us and what experience has taught us is that when females are given the opportunity to take charge of their economic and financial securities, they not only take charge of their own destiny, they transform the trajectories of their families and their communities for the better (Carter 2001). If we may tap into the knowledge females have of needs of their families and assist them to form an economic identity, families will benefit as will communities and nations.

Education should be co-ordinated with opportunities for putting that knowledge to better use. The paper has tried to identify real steps to put in place to realize that dream; it has talked about women entrepreneurship and different challenges and barriers faced by them in establishing new businesses. Women-entrepreneurship is both about the position of women within society as well as their role of entrepreneurship within the similar society. They are faced with particular barriers (like family responsibilities) which need to be overcome for giving them access to the same opportunities as men. The greatest contribution of women in the labour force is a precondition to improve their position in society as well as self-employed women. Policy makers have to promote the networking of associations and foster support and joint ventures amongst national as well as global networks and make possible activities of entrepreneurship by women in the economy.

Another aspect that came to the attention of the authors, especially in developing countries, is the lack of security for women to perform their professional duties. In addition to the above, there are many cases of women harassment in developing countries as compared to the developed countries. This attitude of developing countries towards their women also affects the confidence of their women and results in less women entrepreneurship as women prefer to stay at their homes in these unfavourable conditions. Although there are many laws relating to harassment in these countries, there is a need for proper implementation of these laws by law enforcement agencies. This can help to improve the conditions for women security in these developing countries and result in an increase in women entrepreneurship.

REFERENCES

Carter, S. 2001. Female's Business Ownership: A Review of the Academic, Popular and Internet Literature.
Heemskerk, M. 2003. Self-employment and poverty alleviation: Female's work in artisanal gold mines. *Human Organisation* 62–72.
Myers, S.C. 2005. Corporate financing and investment decisions when firms have information that investors do not have. *Journal of Financial Economics* 187–221.
Storey, D.J. 2009. Entrepreneurship, small and medium sized enterprises and public policy. In Z.J. Acs & D.B. Audretsch (eds), *Handbook of Entrepreneurship Research* 473–511.
Venkataraman, S. 2010. The distinctive domain of entrepreneurship research: An editor's perspective. In J. Katz & R.H.S. Brockhaus (eds), *Advances in Entrepreneurship, Firm Emergence and Growth*. Greenwich, CT: JAI Press.

Knowledge sharing culture in higher education: Critical literature review

Osama F. Ali Al Kurdi
Ahlia University, Kingdom of Bahrain

Ahmad Ghoneim
Brunel University London, UK

Amer Al Roubaie
Ahlia University, Kingdom of Bahrain

ABSTRACT: This paper reviews and analyses the literature on knowledge sharing in a university setting with the aim of identifying and understanding the determinants of knowledge sharing culture, research trends, theories, and future research opportunities for knowledge sharing in higher education institutions (HEIs). Findings suggest that there is disproportionately little knowledge sharing research in HEIs compared to the commercial sector. The review reveals that existing research on HEIs does not consider the determinants of knowledge sharing culture in a comprehensive manner. Research on knowledge sharing in commercial and HEIs in developing economies like Africa, the Middle East and South America is found to be limited. The review shows that future research should consider cultural and behavioural factors at different levels, that is, individual, national, professional teams, language issues and trust that might impact knowledge sharing practices among faculty members in HEIs in developing economies.

1 INTRODUCTION

Knowledge is very important for organizations; many agree that it is one of the most valuable commodities as we emerge to the knowledge base economy (Nonaka & Takeuch 1995, Kukko 2013, Amayah 2013, Fullwood et al. 2013, Goh & Sandhu 2013, Howell & Annansingh 2013). Cabrera & Cabrera (2005) argue that only when knowledge is shared between workers cross the organization, can knowledge management (KM) be effective. Hence, knowledge sharing (KS) is recognized by many as the foundation of knowledge management programmes (Amayah 2013, Fullwood et al. 2013, Kukko 2013). Therefore, creating effective knowledge sharing culture (KSC) is important for the success of KM programmes (Suhaimee et al. 2006). Establishing such a culture is, however, a complex task that is influenced by several individual, social, organizational and technological factors (Riege 2005, Cabrera & Cabrera 2005, Kim & Ju 2008, Kukko 2013). This explains the large number of articles addressing knowledge sharing culture determinants in the business sector literature such as (Gurteen 1999, Cabrera & Cabrera 2005, Reid 2003, Muller 2005, Suhaimee et al. 2006, Michailova & Hutchings 2006, Magnier-Watanabe & Senoo 2010, McAdam et al. 2012) that explore inhibitors to knowledge sharing among employees. Despite the extensive business sector research into knowledge sharing culture factors, there is limited research in the context of Higher Education Institutions (HEIs), which comprehensively examines determinants of knowledge sharing culture influences in developing nations (Kim & Ju 2008, Wang & Noe 2010, Fullwood et al. 2013, Goh & Sandhu 2013, Howell & Annansingh 2013). Due to the importance of knowledge sharing for organizations, particularly HEIs as knowledge intensive organizations, the aim of this paper is to review relevant research on knowledge sharing to identify current state of the art, research focus and future research activities in HEIs. Previous reviews focussed on factors influencing knowledge sharing culture in the commercial sector, and did not comprehensively explore KS practices among faculty members in HEIs.

The paper is organized as follows: section 2 provides an overview of knowledge sharing concepts, and definitions are briefly discussed. The method employed to select the articles for the review is explained. Section 3 subsequently describes the literature related to the research objectives and presents findings. The final section concludes the study and lays out implications and future research areas.

2 KNOWLEDGE, KNOWLEDGE SHARING AND KNOWLEDGE MANAGEMENT: LITERATURE PERSPECTIVE

Despite several attempts to define knowledge, knowledge management or knowledge sharing in the literature, they continue to be debatable topics among academics and practitioners, depending on the context and perspective in which they are used (Nonaka 1994, Davenport & Prusak 1998, Alvi & Leidner 2001, Cabrera & Cabrera 2002, Kakabadse 2003, Wang & Noe 2010, Lloria 2008, Gao et al. 2008, Iqbal & Mahmood 2012, Ragab & Arisha 2013).

Knowledge has been linked in the literature to terms like data, information, experience, intuition, ideas, depending on the context. For example, Nonaka (1994) argues that knowledge is "justified true beliefs" while Alvi & Leinder (1999) define knowledge by distinguishing between data, information and knowledge. Lloria (2008) describes knowledge management as a series of policies and guidelines to facilitate and enable creation, sharing and institutionalize knowledge to achieve a firm's objectives. On the other hand, Gao et al. (2008) define KM as "to manage the activities of the knowledge worker, which is done through facilitating, motivating, leading, and supporting knowledge workers and providing or nurturing a suitable working environment".

Knowledge sharing in the context of work is described as the exchange or dissemination of explicit or tacit data, ideas, experiences, or technology between individuals or group of employees (Cabrera & Cabrera 2002, Wang & Noe 2010). Amayah (2013) concludes that knowledge sharing focusses on the know-how type of knowledge to help others, solve problems, develop new ideas, or implement policies and procedures. Knowledge sharing and knowledge exchange are used interchangeably, however, by Wang & Noe (2010) who state that knowledge exchange includes both, knowledge sharing (knowledge contributor) and knowledge seeking (knowledge searching). On the other hand, knowledge transfer refers only to the movement of knowledge across an organization and not between individuals (cited in Wang & Noe 2010).

3 RESEARCH METHODOLOGY FOR THE LITERATURE REVIEW

For this review the following techniques for systematic literature review process were applied:
Mapping the field of knowledge sharing in HEI through a scoping review

- comprehensive search,
- quality assessment,
- data extraction,
- data synthesis,
- presenting the findings.

A research plan was developed including research questions, publications' inclusion and exclusion criteria, database identification and search key words. The objective of the paper is to determine the current status of knowledge sharing culture research in Higher Education Institutions. The research questions developed to help achieve this objective were:

- What are the key concepts and definitions related to knowledge sharing culture?
- What are the knowledge sharing culture determinants?

- Which determinants of knowledge sharing culture were well researched and which are not?
- What are the main methods used for the studies?
- What are the main findings?

In order to answer the research questions, the inclusion criteria for publications were: publications between 2003 and 2014, English language only, peer reviewed, focus on Higher Education Institutions (public or private), focus on knowledge sharing culture determinates among faculty members, knowledge sharing key concepts, processes and literature review papers. Exclusion criteria were: publications prior to 2003, non-English language, book reviews and chapters and non-academic research.

The ProQuest database was used due its wide coverage of titles, international publishers and comprehensive peer reviewed journals in different areas. Two types of key words were used, general and specific. General keywords aimed to provide comprehensive understanding of knowledge sharing concepts and definitions in general in organization settings, while specific key words aimed to gain current research status in a specific context. General keywords included 'knowledge sharing culture', 'knowledge sharing', and 'knowledge transfer'. Specific keywords included the following: 'knowledge sharing culture' AND 'Higher Education Institutions', 'knowledge sharing' AND 'Higher Education', 'knowledge sharing' AND 'Academics', and 'knowledge sharing' AND 'Faculty'. All keywords had to appear in the document title in order to restrict the search to a manageable number of articles.

The search with general keywords returned 373 publications from 2003 to 2014. To narrow the search to its focus of interest specific key words were applied on the returned articles. This search returned seventy-nine qualitative and quantitative studies, which were subsequently screened by reviewing individual abstracts and applying the inclusion and exclusion criteria. Large numbers of articles were published in the Journal of Knowledge Management, Journal of Knowledge Management and Practice which are among the highest ranked KM journals. Discipline areas included management, human resources management, education management and technology, and information systems.

4 PRESENTATION OF THE FINDINGS

4.1 *Knowledge sharing culture determinants*

In a study by Smith & McKeen (2003), senior knowledge managers described knowledge sharing culture (KSC) as 'where ideas are freely challenged, knowledge learned is applied, and where willingness to share knowledge and teach others is the norm'. Though several definitions describe KM and KS differently, agreement among writers is that both concepts should be treated as a process to manage organizational knowledge and facilitate innovation to compete in the global economy.

In order to promote and encourage a culture of knowledge sharing in organizations, research has focussed on many areas mainly related to organization, people and technology. Technology elements focussed on systems and tools to facilitate sharing, while people and organization elements included culture (that is, national, organizational, individual, team climate), motivations, incentives, trust and individual identity. Knowledge sharing culture determinants will be briefly described in the next sub-sections.

4.1.1 *Technology determinants*

Information technology (IT), information systems (IS) and knowledge management systems (KMS) are widely used; they are described in the literature as knowledge sharing facilitators between employees and a key enabler of KM alongside organization and people (Davenport & Prusak 1998, Alvi & Leinder 1999, Smith & McKeen 2003, Riege 2005, Bock et al. 2005, Berlanga et al. 2008, Seba et al. 2012). Emphasis on the right technology that fits employee needs and promotes all type of communication methods were, however, stressed by Riege (2005), O'Dell & Grayson (1998) and Tsai et al. (2013). The role of IT in promoting knowledge sharing was evident as an enabler in several empirical studies (Kim & Lee 2006, Ahmad & Daghfous 2010, Sharma et al. 2012, Siddique 2012, Kanaan & Gharibeh 2013).

Other studies examined the relationship between IT and other factors (trust and culture) in promoting organizational knowledge sharing (Choi & Lee 2003, Golden & Raghuram 2010, Young et al. 2012, Siddique 2012), and concluded that IT support and infrastructure were less emphasized by workers compared to trust and good knowledge sharing culture. In other words, IT or KMS alone don't achieve effective knowledge sharing in the absence of factors such as trust, culture, organizational climate and leadership support. Systems and technology tools were identified by other studies as barriers to knowledge sharing (Riege 2005, Smith & McKeen 2003), where unrealistic expectations of technology, lack of training on the system, poor usability and design of the system would impede KM and KS efforts. The role of leadership to ensure the proper selection of technology to fit the existing organizational culture was highlighted by Berlanga et al. (2008), Seba et al. (2012) and Tsai et al. (2013).

4.1.2 People and organization determinants

Factors related to people and the organization have dominated knowledge sharing research, some more so than others. In this sub-section, widely cited factors are highlighted.

Organizational culture was the focus of several studies (De Long & Fahey 2000, Li et al. 2006, Al-Alawi et al. 2007, Magnier-Wanatabe & Senoo 2010, Nguyen & Mohamed 2011, Sanz-Valle et al. 2011, Tong et al. 2013). Authors established several dimensions that affect knowledge sharing behaviour including trust, national culture, leadership, organization structure and organizational learning. Sub-cultures, organizational climate, team culture and professional group culture were examined in relation to KS (Ardichvili et al. 2006, King 2008, Chen et al. 2010, Jackson et al. 2010, Magnier-Wanatabe & Senoo 2010, McAdam et al. 2012).

A number of those studies were conducted in Chinese culture where it was found that different levels of culture have direct influence on knowledge sharing behaviour. For example, McAdam et al. (2012) have examined culture role on knowledge sharing processes at different organizational levels in Chinese organizations by developing an integrated cultural framework. They showed that Chinese culture at the corporate, group and individual levels influences knowledge sharing processes. Ardichvili et al. (2006) have examined the impact of national culture factors on knowledge sharing strategies in online communities of practice in three different countries (Brazil, China and Russia). They outlined that KM programmes are influenced by the values and cultural preferences of workers. Li et al. (2006) examined organizational cultural and factors that impact online knowledge sharing between American and Chinese participants in fortune 100 companies. She concluded that sharing knowledge is influenced by national culture differences cross organization and communities of practice (COP).

4.1.3 Behavioural and motivational determinants

In order to encourage knowledge sharing behaviour, many enablers and success factors have been discussed in the literature. Trust interrelation with knowledge sharing culture was the subject of many studies (Alam et al. 2009, Aulawi et al. 2009, Wang & Noe 2010, Casimir et al. 2012, Wickramasinghe & Widyaratne 2012). Rewards (extrinsic and intrinsic), innovation, leadership, incentives, technology, commitment, demographic profiles and job satisfaction were found to influence KS in the business sector (Bock et al. 2005, Aulawi et al. 2009, Arzi et al. 2013, Alam et al. 2009, Wickramasinghe & Widyaratne 2012, von Krogh et al. 2012, Kathiravelu et al. 2013, Tong et al. 2013, Kanaan & Gharibeh 2013). On the other hand, relationships between barriers and KS were identified and examined by Riege (2005), Arntzen & Worasinchai (2012), Sharma et al. (2012), Kukko (2013) and Santos et al. (2012). Findings identified several barriers: lack of time to share knowledge, trust culture, communication media, knowledge sharing culture, training on IT tools, leadership support and commitment, job security, different national culture and unwillingness to use technology. Many papers on KS enablers and barriers were qualitative in nature utilizing survey-based questionnaires and located in western and Asian countries.

The majority of research programmes reviewed in the commercial sector, were conducted in western countries, Malaysia and China. Limited studies were conducted in the Middle East, Africa and South America (Al-Alawi et al. 2007, Alam et al. 2009, Heydari et al. 2011, Seba et al. 2012; Siddique 2012, Kanaan & Gharibeh 2013). Furthermore, the public sector was the topic of a number of studies; comparative papers between the public and private sector's knowledge sharing practices and culture were noticeable as well.

4.2 Knowledge sharing in higher education institutions

Higher education institutions (HEIs) are distinctive from other organizations where knowledge is their input and output (Omerrzel 2011); it is well known that HEIs are in the business of generating and disseminating knowledge (Rowley 2000, Sharimllah et al. 2007, Kim & Ju 2008, Sohail & Daud 2009, Alhammad et al. 2009, Chen et al. 2010, Omerzel 2011, Heydari et al. 2011, Karahoca et al. 2011, Jahani et al. 2011, Sandhu et al. 2011, Nordin et al. 2012, Ramachandran 2013, Howell & Annansingh 2013, Amayah 2013; Goh & Sandhu 2013, Fullwood et al. 2013, Li et al. 2006, Jolaee et al. 2014). There is agreement among authors that HEIs' approach to knowledge management would facilitate the transition to a knowledge-based economy, enhance knowledge sharing, improve educational programmes and consequently improve the overall performance of universities. A university is considered as a platform for academics to share ideas and insights (Martin & Marion 2005). Thus, KM and knowledge sharing should be critical in knowledge intensive organizations such as higher education institutions, where maximizing the intellectual capital allows them to compete in the global market (Swart & Kinnie 2003, Kim & Ju 2008, Sohail & Daud 2009, Karahoca et al. 2011, Fullwood et al. 2013, Goh & Sandhu 2013).

Given the large amount of research focussing on studying knowledge sharing among employees in commercial organizations, one could expect that HEIs have utilized KM and KS strategies applied in the corporate sector. The literature indicates otherwise: there were few attempts by HEIs to implement comprehensive KM and KS programmes (Rowley 2000, Abdullah et al. 2008, Kim & Ju 2008, Chen et al. 2010, Ramachandran 2013, Fullwood et al. 2013, Goh & Sandhu 2013). Additionally, limited research considering knowledge sharing among faculty members within HEIs was found (Kim & Ju 2008, Sohail & Daud 2009, Nordin et al. 2012, Fullwood et al. 2013). The lack of KM and KS application in HEI compared to the business sector can be attributed to the few attempts to utilize the widely recognized benefits of KM (Chen et al. 2010), while unwillingness to share knowledge by faculty can be attributed to lack of systems and policies to protect their intellectual assets (Kim & Ju 2008), the individualistic nature of academics and research (Kim & Ju 2008, cited in Fullwood et al. 2013), the complexity of academic departments (cited in Fullwood et al. 2013) and loyalty to the discipline rather than the organization (cited in Fullwood et al. 2013).

Due to the small number of identified studies on knowledge sharing among faculty members in HEIs, they are subsequently explored in detail with the aim of identifying research trends and studied KS determinants in the following:

- Jolaee et al. (2014) examined factors effecting knowledge sharing intentions among faculty staff in universities utilizing the theory of reasoned action (TRA). A total of 117 usable questionnaires were returned from 200 distributed in one public university in Malaysia. Authors investigated the influence of attitude, subjective norm and trust on knowledge sharing among academics. They concluded that attitudes are positively related to knowledge sharing intention. On the other hand, self-efficacy and subjective norms were not found to effect knowledge sharing intentions. Similarly, trust was not found to impact intention to share knowledge, this finding was conflicting with previous studies finding (Davenport & Prusak 1998, O'Dell & Grayson 1998). Extrinsic rewards were found not to have a positive effect on knowledge sharing among academics; again this was inconsistent with earlier findings. The authors highlighted that these inconsistencies could be contributed to the context of their study (academic staff in a university) while others are in the commercial sector. The authors recommend that their findings should be compared with geographically extended studies and comparison between public and private sector universities in relation to knowledge sharing among academics.

- Fullwood et al. (2013) examined the intentions and behaviour to share knowledge among academics in 11 universities in the UK with 230 valid responses from different disciplines and colleges. Selected universities covered "Post-92" which is teaching led and "Pre-92" which was research driven. A survey-based questionnaire was used to collect data due to the lack of previous research in this area; it covered many factors that affect knowledge sharing, such as intentions and attitudes to knowledge sharing, reward expectations, organizational culture and climate influence, technology platform, leadership, affiliation to institutions and discipline. They showed that knowledge sharing culture among academics in HEIs is individualistic in nature and self-serving; this poses some challenges on HEIs to seek innovative strategies to create, store and disseminate knowledge to students, society and industry. Additionally, responses on the impact of leadership, organization culture and information technology were low and neutral. This finding was not in line with other studies such as that by Wang & Noe (2010). The authors outlined several research gaps such as (1) a lack of measuring the extent of knowledge sharing in universities and impact on organizational success, and (2) the lack of knowledge sharing studies in HEI in different countries, where the role of the national and other types of culture requires further examination.
- Another quantitative study conducted by Howell & Annansingh (2013) aimed to examine path-dependency and cultural influences on knowledge generation and sharing in two UK universities. Two focus groups were used to collect data in the selected universities (Post-1992 and Russell group). The authors found limited knowledge sharing practices in the "Post-1992" university; rather, the sharing took place outside the university, where rewards exist. This finding coincides with other studies by Ipe (2003) and Rowley (2000), where institutional and sub cultures played a key role in determining knowledge sharing behaviour of academic staff.
- Nordin et al. (2012) conducted a study in a Malaysian university to examine knowledge sharing behaviour among academic staff, employing the 'theory of planned behaviour' (TPM) to identify factors influencing an academic's decision to share knowledge and the extent of knowledge sharing. They utilized a structured questionnaire survey to collect data with 187 valid responses. Four factors were considered: attitudes, subjective norm, comply norm, normative norm and 'perceived behaviour control' (PBC). They concluded that only attitudes, comply norm, normative norm, and PBC influence knowledge sharing behaviour among academics. Further research on HEIs in different countries would verify and confirm generalizability of the findings from this study.
- A quantitative study based in the Middle East, Al Husseini & Elbeltagi (2012) investigated the link between knowledge sharing and product and process innovation in six private colleges in Iraq. The authors used a self-administered questionnaire with 230 valid responds. The findings showed that acquiring knowledge (collecting) among faculty members is very high compared to sharing (donating) knowledge. This behaviour varies among departments; academics in one department find it easier to donate knowledge since staff share similar values, interest and beliefs. They add that knowledge donating and collecting are antecedent to product and service innovation. They recommend studying culture as an influencer of knowledge sharing practices in HEI in the Middle East as further research.
- An empirical study conducted by Babalhavaeji & Kermani (2011) studied faculty related factors that influence knowledge sharing in two Iranian universities (public and private). The authors used a survey-based questionnaire with a total of 90 valid responses from teaching staff in the two universities. They found that faculty with an intention to encourage knowledge sharing had a positive attitude towards knowledge sharing culture in HEIs. They showed furthermore that experience of faculty members has a relationship with faculty's knowledge sharing behaviour. Academics with less than five years and more than 20 years experience demonstrated higher sharing behaviour. This finding was contradicted by a study done by Lou et al. (2007) in Thai universities, which explored knowledge sharing behaviour of information management instructors. Lou et al. (2007) outlined that instructors with five to ten years experience are keen to share

knowledge compared to instructors with less than five years of experience. Further research is recommended to study other factors affecting knowledge sharing such as trust, communication and collaboration behaviour among academics.

- Chen et al. (2010) examined knowledge sharing behaviour among academics in a private university in Malaysia. The authors utilized a survey-based questionnaire with 60 valid responses out of 119. Organizational, individual and technological factors were examined; the authors found that incentive systems and personal expectation were major influences of faculty members to share knowledge. Additionally, the study revealed that forcing academics to share knowledge such as research outcome is not as effective as a reward is. Hence, understanding individual factors (internal and external) that prevent knowledge sharing is essential for HEIs. This finding is supported by a study conducted in a business school in India by Basu & Sengupta (2007). The authors call for further research to examine internal and external factors that influence the individual's decision to participate in knowledge sharing activities.
- Additionally, Kim & Ju (2008) examined perception and behaviour factors that influence faculty members to share knowledge across campus in a Korean private university. They utilized a survey-based questionnaire to collect data, and received 78 valid responses out of 109 from different colleges and disciplines. Several potential factors were examined to influence KS in HEI (for example, perceptions, trust, openness, collaboration, reward-system and communication channel). Out of those factors, perceptions and reward systems were found to influence academics to share knowledge in HEIs the most.
- Suhaimee et al. (2006) investigated reward systems as an influence on knowledge sharing culture during KM implementation in Malaysian Public Institutions of Higher Education (PIHE). The authors used a survey-based questionnaire to collect data from 17 PIHE universities targeting IT managers. Findings indicated that incentives, job assessment and promotions were most likely to encourage knowledge sharing behaviour among staff. Additionally, the authors recommended more actions to promote knowledge-sharing culture among staff and academics in PIHE.
- In an earlier study, Dyson (2004) explored knowledge sharing barriers among faculty in an Australian university; the case study targeted the process of creating and sharing knowledge using 25 semi-structured interviews with faculty and students. The study found that lack of time, unwillingness to share knowledge, lack of common cultures and languages were similar to the corporate sector. Each of the findings corresponded to the characteristics of faculty members in universities, for instance, hesitance to share knowledge was associated with the independence nature of faculty members. Lack of culture and language was related to the departmental cultures or disciplines existed in HEIs. Findings pose serious challenges in managing knowledge management process in HEIs, the author concluded.

The table in the Appendix summarizes knowledge sharing determinants explored in the above studies.

4.3 *Discussion*

Based on the reviewed literature on knowledge sharing in the higher education context among faculty members, it can be concluded that an initial understanding of knowledge sharing in higher education has been slowly developing over the past decade. The review also showed, however, that this understanding is fragmented and does not comprehensively consider several factors that might influence faculty to share their knowledge. In the following sub-sections, the focus is on specific areas of knowledge sharing and KSC in HEIs.

4.3.1 *Knowledge management and knowledge sharing definitions*
Certain aspects of knowledge and knowledge management did not enjoy consensus in terms of definitions from academics and practitioners (Wang & Noe 2010, Lloria 2008, Gao et al. 2008, Iqbal & Mahmood 2012, Ragab & Arisha 2013). A widely accepted definition of knowledge in the literature has been put forward by Davenport & Prusak (1998), where knowledge is defined as

"a fluid mix of framed experience, values, contextual information and expert insight that provides a framework for evaluating and incorporating new experiences and information. It organizes and is applied in the minds of 'knowers'. In originations, it often becomes embedded not only in documents or repositories but also in organizational routines, processes, practices and norms". This definition highlighted the complex nature of knowledge; it is a mixture of structured and fluid and comprised of many elements, it exists within people and human complexity. KM on the other hand, had multiple definitions, but Lloria (2008) offers a comprehensive one capturing many KM elements after synthesizing relevant literature and approaches. According to Lloria (2008), KM is a series of policies and guidelines to facilitate and enable creation, sharing and institutionalize knowledge to achieve firm's objectives. Knowledge sharing as a key element of KM had a generally accepted definition in the literature (Cabrera & Cabrera 2002, Wang & Noe 2010) where it was described as the exchange or dissemination of explicit or tacit data, ideas, experiences, or technology between individuals or groups of employees. It is vital for management staff, researchers and industry professionals to carefully examine the contextual elements when setting up KM and KS programmes and strategies to ensure success.

4.3.2 *Knowledge sharing culture*

A positive culture of knowledge sharing is where willingness by workers to share knowledge and teach others is the norm in the organization (Smith & McKeen 2003). Several studies identified KSC as a key factor for successful KM programmes (Suhaimee et al. 2006, Riege 2005, Cabrera & Cabrera 2005). Despite of the importance of ensuring such a culture is in place, it is a complex and multidimensional activity (Kim & Ju 2008, Kukko 2013). The majority of the literature on KSC was found to focus on the commercial sector compared to higher education.

4.3.3 *Knowledge sharing culture determinants*

KSC determinants have been examined and investigated in the literature in relation to technology, people and behaviours and organization (Sharma et al. 2012, Siddique 2012, Kanaan & Gharibeh 2013). Organizational culture, national culture, trust, organizational climate, rewards (extrinsic and intrinsic), innovation, leadership, incentives, commitment, demographic profiles and job satisfaction were found to positively influence KS in the commercial sector among the people and organizational factors examined in the reviewed papers (Alam et al. 2009, Aulawi et al. 2009, Wang & Noe 2010, Casimir et al. 2012, Bock et al. 2005, Arzi et al. 2013, Wickramasinghe and Widyaratne 2012, von Krogh et al. 2012, Kathiravelu et al. 2013, Tong et al. 2013, Kanaan & Gharibeh 2013). Despite the well-researched determinants of KSC, the comprehensive examination of joint factors such as national culture, sub-cultures, power perspective, interpersonal sharing *versus* technology-aided sharing, perspectives in developing economies require further investigation.

4.3.4 *Knowledge sharing culture in HEIs*

The majority of the reviewed studies on knowledge sharing in an HEI context investigated the following determinates to have relationship with knowledge sharing behaviour and intention: trust, technology, rewards, organizational climate, incentives, subjective norm, and attitudes (Jolaee et al. 2014, Fullwood et al. 2013, Nordin et al. 2012, Babalhavaeji & Kermani 2011, Chen et al. 2010, Kim & Ju 2008, Suhaimee et al. 2006).

The reviewed literature does not consider the determinants affecting knowledge sharing culture and practices in HEIs in a comprehensive manner. Whilst these have been well researched to a certain degree in the corporate sector, the relationship between determinants and influences on knowledge sharing in HEIs needs further research. Cultural factors such as national culture, organizational climate, academic culture, religion, sub- and team cultures, language and gender) would impact academics' decisions to participate in KM and knowledge sharing activities (Fullwood et al. 2013, Al Husseini & Elbeltagi 2012, Nordin et al. 2012, Dyson 2004). Other factors such as knowledge communication methods, trust, internal and external influences of KS in HEIs need to be explored (Babalhavaeji & Kermani 2011, Cheng et al. 2009). An in-depth study of factors influencing knowledge sharing among faculty members in HEIs would assist universities to adopt appropriate strategies to manage their intellectual assets, enhance performance, research output and teaching activities.

5 CONCLUSIONS

This paper has reviewed existing empirical studies on knowledge sharing culture in HEIs. In a global knowledge driven economy and increased competition, organizations are continuously looking to effectively manage organizational knowledge to stay competitive. HEIs are in the business of creating and disseminating knowledge; therefore, creating a knowledge sharing culture is essential for universities' continuous improvements in research output and knowledge transfer to society.

The review revealed that knowledge sharing culture studies in HEIs is still very limited and developing. It can thus be concluded that the existing empirical literature provides a fragmented insight into knowledge sharing culture in HEIs. The main research objectives focussed on investigating perceptions and behaviour of faculty members towards sharing knowledge among colleagues. The key knowledge sharing culture determinants identified were related to people, organization and technology. A limited number of variables in each area were comprehensively examined in relation to knowledge sharing culture. Further determinants such as culture, national culture, sub-cultures, language, trust, and communication methods offer additional research avenues in HEIs for researchers.

The authors have systematically reviewed relevant studies of knowledge sharing culture focussing on the higher education context, specifically among faculty members. The review of extant empirical studies has revealed the following research implications and opportunities for further research:

- Several studies in the business sector focussed on the influence of technology on knowledge sharing practices; this included explicit knowledge since it is easier to codify. It is widely accepted in the literature that tacit knowledge is exchanged or transferred using face-to-face communication; future research might compare using technology to share knowledge through KMS and face-to-face.
- Most of the non-western culture studies of HEIs were carried out in Malaysia; more studies are needed to understand how different cultures would impact knowledge sharing practices in countries in Africa, the Middle East and South America and compare findings with other regions.
- Case studies focussing on knowledge sharing culture are needed; most of the HEIs literature reviewed utilized self-completion survey questionnaires, this method might be influenced by respondents' attitudes and beliefs toward knowledge sharing.
- More longitudinal studies focussing on factors influencing employees to share knowledge are needed; this type of study ensures consistency of findings and measures behaviour of staff before and after implementation of KM programmes.
- Studies focussing on the impact of knowledge sharing culture and practices on HEIs performance are needed; performance measures such as research output, growth, innovation, ranking and reputation need further study. This will demonstrate the value of KM and KS initiatives and link them to strategic performance of HEIs.
- The majority of studies reviewed on HEIs focussed on views from academics; additional views from senior management, administrators and government officials need to be considered.

REFERENCES

Abdullah, R., Selamat, M., Jaafar, A., Abdullah, S. & Sura, S. 2008. An Empirical Study of Knowledge Management System Implementation in Public Higher Learning Institution 8(1). Retrieved in February 2014 from http://citeseerx.ist.psu.edu/viewdoc/summary?doi=10.1.1.131.8993.

Ahmad, N. & Daghfous, A. 2010. Knowledge sharing through inter-organizational knowledge networks: Challenges and opportunities in the United Arab Emirates. *European Business Review* 22(2): 153–174.

Al-Alawi, A.I., Al-Marzooqi, N.Y. & Mohammed, Y.F. 2007. Organizational culture and knowledge sharing: Critical success factors. *Journal of Knowledge Management* 11(2): 22–42.

Alam, S., Abdullah, Z., Ishak, N. & Zain, Z. 2009. Assessing knowledge sharing behaviour among employees in SMEs: An empirical study. *Business Research* 1998: 115–122.

Alavi, M. & Leidner, D.E. 1999. Knowledge management system: Issues, challenges, and benefits. *Communications of the AIS* 1(7): 1–37.

Alhammad, F., Faori, S. Al & Husan, L.A. 2009. Knowledge sharing in Jordanian Universities. *Journal of Knowledge Management* 10(3).

Al Husseini, S. & Elbeltagi, I. 2012. *Knowledge Sharing and Innovation: An Empirical Study in Iraqi Private Higher Education Institutions*.

Amayah, A.T. 2013. Determinants of knowledge sharing in a public sector organization. *Journal of Knowledge Management* 17(3): 454–471.

Ardichvili, A., Maurer, M., Li, W., Wentling, T. & Stuedemann, R. 2006. Cultural influences on knowledge sharing through online communities of practice. *Journal of Knowledge Management* 10(1): 94–107.

Arntzen, A. & Worasinchai, L. 2012. Analysis of the barriers of knowledge sharing: An insight of Thai firms. *Conference on Knowledge*. Retrieved in February 2014 from http://books.google.com/books?hl=en&lr=&id=ReFcQ4vjHAkC&oi=fnd&pg=PA57&dq=Analysis+of+the+Barriers+of+Knowledge+Sharing+:+An+Insight+of+Thai+Firms&ots=Oo2GhUyUlf&sig=VZc__6qhL0Rd8pOvBd_dF9GHZVc.

Arzi, S., Rabanifard, N., Nassajtarshizi, S. & Omran, N. 2013. Relationship among reward system, knowledge sharing and innovation performance. *Interdisciplinary Journal of Contemporary Research in Business* 5(6): 115–142.

Aulawi, H., Sudirman, I., Suryadi, K. & Govindaraju, R. 2009. Literature review towards knowledge enablers which is assumed significantly influences KS Behavior. *Journal of Applied Sciences* 5: 2262–2270.

Babalhavaeji, F. & Kermani, Z. 2011. Knowledge sharing behaviour influences: A case of library and information science faculties in Iran. *Library and Information Science* 16(1): 1–14. Retrieved from http://biomed2011.um.edu.my/filebank/published_article/2815/article-1.pdf

Basu, B. & Sengupta, K. 2007. Assessing success factors of knowledge management initiatives of academic institutions: A case of an Indian business school. *Electronic Journal of Knowledge Management* 5(3): 273–282.

Berlanga, A.J., Sloep, P.B., Kester, L., Brouns, F., van Rosmalen, P. & Koper, R. 2008. Ad hoc transient communities: Towards fostering knowledge sharing in learning networks. *International Journal of Learning Technology* 3(4): 443.

Bock, G.-W., Zmud, R.W., Kim, Y.-G. & Lee, J.-N. 2005. Behavioral intention formation in knowledge sharing: Examining the roles of extrinsic motivators, social-psychological forces, and organizational climate. *MIS Quarterly* 29(1): 87–111.

Cabrera, A. & Cabrera, E.F. 2002. Knowledge-sharing dilemmas. *Organization Studies* 23(5): 687–710.

Cabrera, E.F. & Cabrera, A. 2005. Fostering knowledge sharing through people management practices. *International Journal of Human Resource Management* 16(5): 720–735.

Casimir, G., Lee, K. & Loon, M. 2012. Knowledge sharing: Influences of trust, commitment and cost. *Journal of Knowledge Management* 16(5): 740–753.

Chen, J., Sun, P.Y.T. & McQueen, R.J. 2010. The impact of national cultures on structured knowledge transfer. *Journal of Knowledge Management* 14(2): 228–242.

Choi, B. & Lee, B. 2003. An empirical investigation of KM styles and their effect on corporate performance. *Journal of Information and Management* 40(5): 403–417.

Davenport, T.H. & Prusak, L. 1998. *Working Knowledge: How Organizations Manage What They Know*. Boston, MA: Harvard Business School Press.

De Long, D.W. & Fahey, L. 2000. Diagnosing cultural barriers to knowledge management. *The Academy of Management Executive*. Retrieved in February 2014 from http://amp.aom.org/content/14/4/113.short.

Dyson, L. 2004. Barriers to sharing and creating knowledge in higher education. Retrieved from http://epress.lib.uts.edu.au/research/handle/10453/1897 in Feb. 2014.

Fullwood, R., Rowley, J. & Delbridge, R. 2013. Knowledge sharing amongst academics in UK universities. *Journal of Knowledge Management* 17(1): 123–136.

Gao, F., Li, M. & Clarke, S. 2008. Knowledge, management and knowledge management in business operations. *Journal of Knowledge Management* 12(2): 3–17.

Goh, S. & Sandhu, M. 2013. Knowledge sharing among Malaysian academics: Influence of affective commitment and trust. *Electronic Journal of Knowledge Management* 11(1): 38–48.

Golden, T.D. & Raghuram, S. 2010. Teleworker knowledge sharing and the role of altered relational and technological interactions. *Journal of Organizational Behavior* 31(8): 1061–1085.

Gurteen, D. 1999. Creating a knowledge sharing culture. *Knowledge Management Magazine* 1–4. Retrieved from http://www.providersedge.com/docs/km_articles/Creating_a_K-Sharing_Culture_-_Gurteen.pdf in February 2014.

Heydari, A., Armesh, H., Behjatie, S. & Manafi, M. 2011. Determinant of incentive factors in knowledge sharing. *Interdisciplinary Journal of Contemporary Research in Business* 2005: 83–96.

Howell, K.E. & Annansingh, F. 2013. Knowledge generation and sharing in UK universities: A tale of two cultures? *International Journal of Information Management* 33(1): 32–39.

Ipe, M. 2003. Knowledge sharing in organisations: A conceptual framework. *Human Resource Development Review* 2(4): 337–359.

Iqbal, J. & Mahmood, Y. 2012. Reviewing Knowledge Management Literature 1005–1027. Retrieved in February 2014 from http://journal archives24.webs.com/1005–1026.pdf.

Jackson, T., Parboteeah, P. & Morgan, V. 2010. The Role of National Culture in Knowledge Sharing: A Multinational Corporation Perspective.

Jahani, S., Ramayah, T. & Effendi, A.A. 2011. Is reward system and leadership important in knowledge sharing among academics? *American Journal of Economics and Business Administration* 3(1): 87–94.

Jolaee, A., Nor, K.M., Khani, N. & Yusoff, R.M. 2014. Factors affecting knowledge sharing intention among academic staff. *International Journal of Educational Management* 28(4): 413–431.

Kakabadse, N. 2003. Reviewing the knowledge management literature: Towards a taxonomy of knowledge management. Retrieved in February 2014 from http://www.emeraldinsight.com/journals.htm?articleid=1506523&show=abstract.

Kanaan, R. & Gharibeh, A. 2013. The impact of knowledge sharing enablers on knowledge sharing capability: An empirical study on Jordanian telecommunication firms. *European Scientific Journal* 9(22): 237–258. Retrieved from http://www.eujournal.org/index.php/esj/article/view/1651.

Karahoca, A., Kanbul, S. & Laal, M. 2011. Knowledge management in higher education. *Procedia Computer Science* 3: 544–549.

Kathiravelu, S., Mansor, N. & Kenny, K. 2013. Factors influencing knowledge sharing behavior (KSB) among employees of public services in Malaysia. *Hrmars.com* 2(3): 107–120.

Kim, S. & Ju, B. 2008. An analysis of faculty perceptions: Attitudes toward knowledge sharing and collaboration in an academic institution. *Library and Information Science Research* 30(4): 282–290.

Kim, S. & Lee, H. 2006. The impact of organizational context and information technology. *Public Administration Review* 370.

King, W.R. 2008. Questioning conventional wisdom: Culture-knowledge management relationships. *Journal of Knowledge Management* 12(3): 35–47.

Kukko, M. 2013. Knowledge sharing barriers in organic growth: A case study from a software company. *Journal of High Technology Management Research* 24(1): 18–29.

Li, Z., Yezhuang, T. & Zhongying, Q. 2006. The impact of organizational culture and knowledge management on organizational performance. *Proceedings of the Information Resources Management Association*.

Lloria, B.M. 2008. A review of the main approaches to knowledge management. *Knowledge Management Research and Practice* 6(1): 77–89.

Lou, S.J., Yang, Y.S. & Shih, R.C. 2007. A study on the knowledge sharing behaviour of information management instructors at technological universities in Taiwan. *World Transactions on Engineering and Technology Education* 6(1): 143–149.

McAdam, R., Moffett, S. & Peng, J. 2012. Knowledge sharing in Chinese service organizations: A multi-case cultural perspective. *Journal of Knowledge Management* 16(1): 129–147.

Magnier-Watanabe, R. & Senoo, D. 2010. Shaping knowledge management: Organization and national culture. *Journal of Knowledge Management* 14(2): 214–227.

Martin, J.S. & Marion, R. 2005. Higher education leadership roles in knowledge processing. *The Learning Organization* 12(2): 140–151.

Michailova, S. & Hutchings, K. 2006. National Cultural Influences on Knowledge Sharing: A Comparison of China and Russia. *Journal of Management Studies* May.

Muller, R. 2005. The influence of incentives and culture on knowledge sharing. Retrieved from http://ieeexplore.ieee.org/xpls/abs_all.jsp?arnumber=1385745 in February 2014.

Nguyen, H.N. & Mohamed, S. 2011. Leadership behaviors, organizational culture and knowledge management practices: An empirical investigation. *Journal of Management Development* 30(2): 206–221.

Nonaka, I. 1994. A dynamic theory of organizational knowledge creation. *Organizational Science* 5(1): 14–37.

Nonaka, I. & Takeuchi, H. 1995. *The Knowledge Creating Company*. Oxford: Oxford University Press.

Nordin, N., Daud, N. & Osman, W. 2012. Knowledge sharing behaviour among academic staff at a public higher education institution in Malaysia. *waset.org* 234–239.

O'Dell, C. & Grayson, C.J. 1998. If only we knew what we know: Identification and transfer of internal best practices. *California Management Review* 40(3): 154–174.

Omerzel, D. 2011. Knowledge management and organisational culture in higher education institutions. *European Management* 111–140.

Ragab, M.A.F. & Arisha, A. 20130. Knowledge management and measurement: A critical review. *Journal of Knowledge Management* 17(6): 873–901.

Ramachandran, S.D. 2013. Knowledge management practices and enablers in public universities: A gap analysis. *Campus-wide Information Systems* 30(2): 76–94.

Reid, F. 2003. Creating a knowledge-sharing culture among diverse business units. *Employment Relations Today* 30(3): 43–49.

Riege, A. 2005. Three-dozen knowledge-sharing barriers managers must consider. *Journal of knowledge Management*.

Rowley, J. 2000. Is higher education ready for knowledge management? *International Journal of Educational Management* 14(7): 325–333.

Sandhu, M.S., Jain, K.K. & Ahmad, I.U.K.B. 2011. Knowledge sharing among public sector employees: Evidence from Malaysia. *International Journal of Public Sector Management* 24(3): 206–226.

Santos, V.R., Soares, A.L. & Carvalho, J.A. 2012. Case study knowledge sharing barriers in complex research and development projects: An exploratory study on the perceptions of project managers. *Knowledge and Process Management* 19(1): 27–38.

Sanz-Valle, R., Naranjo-Valencia, J.C., Jiménez-Jiménez, D. & Perez-Caballero, L. 2011. Linking organizational learning with technical innovation and organizational culture. *Journal of Knowledge Management* 15(6): 997–1015.

Seba, I., Rowley, J. & Delbridge, R. 2012. Knowledge sharing in the Dubai Police Force. *Journal of Knowledge Management* 16(1): 114–128.

Sharimllah Devi, R., Chong, S.C. & Lin, B. 2007. Organisational culture and KM practices from the perspective of institutions of higher learning. *International Journal of Management in Education* 1(1/2): 57–79.

Sharma, B., Singh, M. & Neha. 2012. Knowledge sharing barriers: An approach of interpretive structural modeling. *IUP Journal of Knowledge* 35–53.

Siddique, C.M. 2012. Knowledge management initiatives in the United Arab Emirates: A baseline study. *Journal of Knowledge Management* 16(5): 702–723.

Smith, H.A. & McKeen, J.D. 2003. Instilling a knowledge-sharing culture. *Queen's Centre for Knowledge-based Enterprises* 20(1): 1–17.

Sohail, M.S. & Daud, S. 2009. Knowledge sharing in higher education institutions: Perspectives from Malaysia. *Vine* 39(2): 125–142.

Suhaimee, S., Bakar, A., Zaki, A. & Alias, R. 2006. Knowledge sharing culture in Malaysian public institutions of higher education: An overview. 354–359.

Swart, J. & Kinnie, N. 2003. Sharing knowledge in knowledge-intensive firms. *Human Resource Management Journal* 13(2): 60–75.

Tong, C., Wah Tak, W.I. & Wong, A. 2013. The impact of knowledge sharing on the relationship between organizational culture and job satisfaction: The perception of information communication and technology (ICT) practitioners in Hong Kong. *International Journal of Human Resource Studies* 3(1): 9.

Tsai, M., Chang, H., Cheng, N. & Lien, C. 2013. Understanding IT professionals' knowledge sharing intention through KMS: A social exchange perspective. *Quality and Quantity* 47(5): 2739–2753.

von Krogh, G., Nonaka, I. & Rechsteiner, L. 2012. Leadership in organizational knowledge creation: A review and framework. *Journal of Management Studies* 49(1): 240–277.

Wang, S. & Noe, R.A. 2010. Knowledge sharing: A review and directions for future research. *Human Resource Management Review* 20(2): 115–131.

Wickramasinghe, V. & Widyaratne, R. 2012. Effects of interpersonal trust, team leader support, rewards, and knowledge sharing mechanisms on knowledge sharing in project teams. *Vine* 42(2): 214–236.

Young, M.-L., Kuo, F.-Y. & Myers, M.D. 2012. To share or not to share: A critical research perspective on knowledge management systems. *European Journal of Information Systems* 21(5): 496–511.

Appendix 1. Summary of research into knowledge sharing (KS) in higher education.

No	Author(s) and year	Country	Methodology	Sample	Determinants researched	Relevant findings
1	Jolaee et al. (2014)	Malaysia	Survey-based questionnaire	117	• Attitudes • Subjective norm • Trust	• Attitudes are positively related to knowledge sharing intention • Self-efficacy and subjective norms were not found to effect knowledge sharing intentions; trust was not found to impact intention to share knowledge
2	Fullwood et al. (2013)	UK	Survey	230	• Intention to share • Types of knowledge shared • Organization climate • Rewards	• Knowledge sharing culture is individual in nature and self-serving in universities • Leadership, organization culture and information technology had low impact on KS behaviour
3	Howell & Annansingh (2013)	UK	Focus groups	2 focus groups	• Organizational culture • Sub-cultures • Path-dependency	• Limited knowledge sharing practices in the "Post 1992" universities • Institutional sub-cultures plays key role in sharing knowledge
4	Nordin et al. (2012)	Malaysia	Structured questionnaire survey	187	• Attitudes towards KS • Subjective norms • Comply norms • Normative norm	• Only attitudes, comply norm, normative norm, and PBC have influenced knowledge sharing behaviour among academics.
5	Al Husseini & Elbeltagi (2012)	Iraq	Self-administered questionnaire	230	• Relationship between knowledge sharing and process innovation in HEIs	• Knowledge sharing intention among faculty is lower than knowledge collecting • Departmental culture impact knowledge sharing behaviours
6	Babalhavaeji & Kermani (2011)	Iran	Survey-based questionnaire	90	• Attitudes • Intention to share knowledge • Intrinsic motivation • Length of experience	• Faculty with higher experience years tend to share knowledge more than lower experience years
7	Chen et al. (2010)	Malaysia	Survey-based questionnaire	60	• Organizational factors • Individual factors • Technological factors	• Forcing academics to share knowledge such as research outcome not effective as a reward • Understanding individual factors (internal and external) that prevent knowledge sharing is essential for HEI
8	Kim & Ju (2008)	Korea	Survey-based questionnaire	78	• Trust • Collaboration • Openness to share • Reward system	• Trust and reward system found to highly influence faculty members decision to share knowledge
9	Suhaimee et al. (2006)	Malaysia	Survey-based questionnaire	17	• Incentives • Promotions • Job assessments	• Knowledge sharing culture is positively influenced by incentives, promotions and job assessments
10	Dyson (2004)	Australia	Case study	25 semi-structured interviews	• Barriers to sharing knowledge among faculty members	• Lack of time and unwillingness to share were found to prevent KS among faculties • lack of common culture and language found to negatively impact KS

The Bahraini corporate governance code: Its effect on the corporate sector

Salah H. Al Hasan
Ahlia University, Kingdom of Bahrain

ABSTRACT: The purpose of the Bahraini corporate governance code is to put the principles of best corporate governance practices into place, and to provide protection to the stakeholders of the company and its investors through compliance with those principles. International experience has shown that the results of good corporate governance are enhanced value of the companies, protecting the investors and attracting investments. This paper examines two empirical studies. First, it examines the relation between the corporate governance characteristics including financial expertise of the board, independent members' composition, frequency of the meeting of committee of audit and boards of directors, board size and the quality of the audit. Secondly, the study examines the efficiency of higher-quality auditors and corporate governance characteristics in constraining earnings management.

1 INTRODUCTION

The issues of quality of audit, corporate governance (CG), and earnings management have received considerable attention from the auditing profession, government regulators, and the public, especially after the recent high profile corporate scandals, the results of which have renewed the importance of an independent audit with linkage to the CG monitoring role (DeFond et al. 2005). Agency theory provides an explanation for why an independent audit is important to the financial market. Wallace (1980) and Lin & Hwang (2010) state that an independent audit helps to reduce the agent-principal conflict by providing assurances that the financial statements are prepared carefully and free from any mistake. It also reduces the likelihood of illegal reporting practices and accounting fraud, such as earnings management, so that market participants can use financial reports without any hesitation (Wallace 1980). Moreover, the auditors can be considered as part of the CG's structure because they regulate the quality of the process of the financial reporting (Beasley & Salterio 2001). The auditors can improve the financial reporting quality through their willingness and competence to report accounting misstatement (DeAngelo 1981) and to respond to aggressive earnings conservatism (Ruddock et al. 2006).

Generally shareholders rely on the ability of the boards and their committees to control the independence of both auditors and management. Thus, the responsibility for the quality of financial reporting is laid on the effectiveness of boards and their committees. The majority of prior studies have concentrated on the committee of audit role as the main factor to ensure financial information integrity and handling with issues related to external audit (Bedard & Gendron 2010, Chen et al. 2005, Abbott et al. 2000). Given that the responsibility for nominating and removing external auditors falls to the board of directors and the members of the committee of audit, their duty is, however, equally important in promoting a higher level of financial reporting quality.

Likewise, some studies have suggested that the effectiveness of committee of audit is linked with whole board composition (Boo & Sharma 2008, Cohen et al. 2002, Collier & Gregory 1996, Menon & Williams 1994). Thus, in this paper, while the demand for a higher quality auditor is recognized, the supervision of roles of the committee of audit and board are argued to be the more important mechanisms through which to promote a higher quality of financial reporting.

Table 1. Listed companies.

Industry group	Number	Suspended
Commercial banks	7	–
Investments	12	2
Services	9	–
Insurance	5	–
Industrial	3	–
Hotels and tourism	5	1
Preferred share	1	–
Closed companies	2	–
Non-Bahraini companies	4	2
Total	48	5

Source: Bahrain Bourse web-site [http://www.bahrainbourse.net/bhb/market.asp?page=market&sec=BOD_Meeting_AGM]

Specifically, this paper examines the relation between the CG characteristics relating to the financial expertise of the board, independent members composition, frequency of the meeting of committee of audit and boards of directors and board size, and the quality of the audit in respect to constraining earnings management. It has been claimed that firms that have effective boards and audit committees are constantly demanding auditors of higher quality because, by employing auditors of higher quality, they add credibility to financial reports, increase the value of the firm and are able to protect themselves from damage to their names and exposure to legal actions; in their entirety these factors promote the interests of shareholders (Carcello et al. 2002).

If there is a lack of participants in the market to supervise the earnings report, they could predict that the firm's directors are in strong supervision to have less in manipulation of earnings. Thus, this paper claims that firms that have supervision mechanisms which consist of the committee of audit and the board of directors with effective characteristics and higher quality auditor are expected to have a higher ability to constrain opportunistic earnings.

2 CONTEXT: CORPORATE GOVERNANCE IN BAHRAIN

Generally, many of the CG principles are embedded in the regulations and rules in force in Bahrain, which pertains to the businesses and activities of the economy under the provisions of Commercial Companies Law – Article 4, "Any type of commercial company based in the Kingdom of Bahrain, must be subject to this law's provisions". There are provisions of CG, most notably in the Law of the Commercial Companies and its implementation, the regulations and the guide of the Central Bank of Bahrain (CBB), and the law establishing and organizing the Bahrain Bourse which represents the largest 48 listed companies and divided into nine sectors, in which five companies are presently suspended from operating (see Table 1). Listed companies are classified according to their activity and the Bahrain Bourse operates as an independent institution supervised by an independent board, chaired by the Governor of the Central Bank of Bahrain.

From the late 1990s until the second quarter of 2008, the economy of Bahrain has observed growth in investment as result of the constant increase in the prices of oil and the improvement of the political and legal frameworks. New regulations and laws have been issued and modified to protect existing stakeholders and investors, and to encourage national and foreign investment in Bahrain. Improving practice of CG also can help the Kingdom of Bahrain to attract more investment. In year 2008, Bahrain was one of the top 20 free economies in the world in terms of index of economic freedom. Furthermore, Bahrain's companies can improve their growth and performance by taking other measures to support investment, such as ensuring higher accountability and transparency in the process of decision making, and creating strong boards of directors. The majority of businesses in Bahrain have taken the form of companies' limited liability or private Bahraini Joint Stock Closed Company (BSCC) for services that are required by rules and regulations to be provided by a closed joint stock company.

Also it is been observed that many businesses have changed the general partnerships to the closed joint stock companies or limited liability companies. Moreover, Bahrain has a large number of companies in the market, large number of banks in the financial institutions, and large number of medium and small sized companies. It is also noted that foreigners may own up to 100% of the activities of the business of many companies and services. Citizens of the Gulf Co-operation Council (GCC) are provided a national treatment with few exceptions. The Government of Bahrain has established Bahrain Mumtalakat Holding Company (Mumtalakat) that operates on lines of commercial and entrusted to the ownership of several assets. At the end of 2010, the Kingdom of Bahrain adopted the new CG code, because Bahrain was looking forward to increasing the transparency and the market value.

3 OBJECTIVES OF THE RESEARCH

The research associated with this paper has the following objectives

1. To examine the role of external audit and the mechanisms of corporate governance (CG) in constraining earnings management in the Kingdom of Bahrain.
2. To identify the relationship between the committee of audit, the board of directors, and audit quality.

The aim of the associated Ph.D. thesis is to examine the effect of the roles of the audit committee, board of directors, and external audit on the quality of financial reporting (including, quality of audit and management of earnings), for the study's purpose, CG is observed as a system of balances and checks to protect the interests of shareholders. Furthermore, the research focusses on financial statements.

A question answered in the research is "Do the external audit and mechanisms of CG constrain the practices of the management of earnings in the Kingdom of Bahrain?"

4 METHODOLOGY

Corporate governance (CG) is a new phenomenon in the Kingdom of Bahrain and there has been no research on the quality audit, CG and earnings management in Bahrain. Therefore, this study is looking into the relationship between CG and the audit quality and between CG and auditor quality in respect of constraining earnings management in Bahraini firms. Moreover, the present research will collect the secondary data from Bahraini listed companies. Unluckily, there are no data services available in the Kingdom of Bahrain. Therefore, this study is based on data collected through the analysis of the annual reports and a designed questionnaire of the listed companies in Bahrain.

5 LITERATURE REVIEW

The study examines the relationship between three elements. The first is the characteristics of CG, the second element is the auditor quality includes three proxies (non-audit service fees, audit fees, and the engagement of industry-specialist auditors) and thirdly, earnings management. Shleifer & Vishny (1997) defined the CG field as a study of the processes of ensuring investments profitability, reducing the financial investors, and methods that resource suppliers. Charreaux (1997) believes that the system of CG includes all the mechanisms that regulate the behaviour of the managers and the demarcation of their own decision. Michel Albert considers that CG gives managers a unique goal: to maximize dividends and profits (Albert 1994).

Although it is admitted that what constitutes CG is still a matter of debate (Cadbury 2000), it significantly analyses the issue from the viewpoints of both public policy perspective and the perspective of the corporation. From the perspective of the corporation, the emerging consensus is that CG is about maximizing the value of the subject to meet financial needs of a corporation, legal obligations and other contractual matters. This comprehensive definition stresses the need for boards of directors to achieve a balance between the shareholders' interests with the other stakeholders and investors in order to achieve sustainable value over the long term.

Also CG expects, through the audit committees, to improve audit quality. DeAngelo (1981) and Watts & Zimmerman (1986) defined the second element (audit quality) as the auditors' competence to prevent or detect errors and objectivity (in fact mind and appearance) of auditors in reporting such errors. The terms "auditor quality" and "audit quality" are supposed to be synonymous, and this is in line with the recommendation of Clarkson & Simunic (1994) that "the audit of the firm is supposed to supply a single level of audit quality at a moment in time". This study will use three proxies to measure the auditor quality including non-audit service fees, audit fees, and the engagement of industry-specialist auditors.

In addition, the code of CG of Bahrain requires that the committee of audit must possess high levels of competence and integrity. It is responsible for reviewing the financial report and oversees the independence of the external auditors (Hasan & Ahmed 2012). Audit committee may appear in some studies to be related to the management of earnings by using different constructs of the effectiveness of the audit committee, such as a member of a committee with financial expertise (Kalbers & Fogarty 1993), size of the board (Yermack 1996, Xie et al. 2003), independent directors with financial motivation (Chtourou et al. 2001a,b), board size (Yermack 1996, Xie et al. 2003), and independence and composition (Klein 2002).

Moreover, the aim of this research is to examine the relation between the audit committee, board of directors, and auditor quality in constraining earnings management. There is no single definition of the term earnings management. Schipper (1989) defined the management of earnings as "the purposeful mediation in the process of the external financial reporting, in order to get some private gain". Healy & Wahlen (1999) claim that management of earnings occurs when managers use their opinion when preparing the financial statements with the purpose of not reporting on the firm's actual economic performance or in order to gain the benefit of the adjusted figure.

Consistent with this description and definition of the benefit of the study of management of earnings views as opportunistic behaviour of management. Managers involve with the opportunistic earnings for various reasons, such as gaining compensation bonus (Healy 1985, Holthausen et al. 1995, Gaver at al. 1995), the avoidance of debt covenant violation (DeFond & Jiambalvo 1994, Sweeney 1994), prevention of decreases of earnings and losses (Burgstahler & Dichev 1997, Barth et al. 1999), and compensation for political or regulatory costs (Jones 1991, Cahan 1992, Han & Wang 1998).

The following points will, however, explain in brief about the effectiveness of the board of directors.

5.1 Size of the board

Lipton & Lorsch (1992) recommend that the size of the board must not be more than eight or nine directors. Jensen (1993) argues that when the board has more than seven or eight members, it is less effective due to the problems to the process of co-ordination, sequentially, contribute lack supervision. In other words, smaller boards are more effective as directors are able to communicate with each other and the meeting is easier to control.

5.2 Independence of directors

Directors are responsible for monitoring managers and controlling the management of the company day after day (Fama 1980, Fama & Jensen 1983, Brennan & McDermott 2004). Therefore, higher numbers of independent directors on boards are expected to monitor functions more effectively, which then leads to financial statements being more reliable.

Also it is been found that independent directors can develop their reputations in making decisions (Fama & Jensen 1983). The study by Beasley (1996) indicates that the largest rate of independent directors on the boards result in a negative impact on financial statement fraud. Regarding earnings management, a stream of literature on independent boards and management of earnings indicate that firms that have a higher percentage of independent members on the board faced a lower incidence of mismanagement of earnings (Klein 2002, Xie et al. 2003, Davidson et al. 2005, Peasnell et al. 2005). In brief, all of these studies recognize that an independent board facilitates effective monitoring.

5.3 Financial expertise

The experience and knowledge of the board are important elements in confirming the effectiveness of the supervision functions of the board. Carcello et al. (2002) suggest that the members of the board with experience of a higher number of positions are more demanding of high audit quality work. Moreover, Chtourou et al. (2001) argue that directors with experience are less likely to be connected with management of earnings. The conclusion of both studies is that levels of high of expertise in board members lead to higher incentive to monitoring. Xie et al. (2003) found that earnings management occurred less in firms that are controlled by a board with high financial and corporate background. In brief, all of these studies acknowledge that the boards who have specific experience and knowledge are useful in supervising the management.

5.4 Board meetings

A director is responsible for attending board meetings and responsible for taking decisions that are made in the meeting (Ronen & Yaari 2008). Conger et al. (1998) indicate that more regular meetings of a board can improve its effectiveness. The meetings are the main operations of the board (Vafeas 1999) and signs of the efforts that have been made by the directors (Ronen & Yaari 2008).

Busy boards that meet more often are more likely to manage their responsibilities in accordance with the shareholders' interests (Vafeas 1999) because more meeting-time can be helpful in controlling issues such as management of earnings and conflicts of interest (Habbash et al. 2010), and in putting more effort into monitoring the integrity of financial reporting and improving the audit quality. A study was conducted by Xie et al. (2003), employing a sample of 282 observations firm year; they point out that the board that meets frequently may have time to look at issues such as earnings management. Their results conclude that the management of earnings is significantly negatively associated with the number of the meetings of the board.

The studies in the previous sub-sections found that boards more independent directors who are equipped with financial expertise, are smaller in size, and meet more often, are effective in their role of supervision. Similarly, the following sub-sections will explain in brief the effectiveness of the committee of audit.

5.5 Size of committee of audit

The size of a committee of audit varies and it depends on the needs of the company and the extent of the responsibilities delegated to the committee. According to the UK CG Code (2010), "The board should be composed of the committee of audit of at least three members, most of [whom] should be non-executive and independent, and that the chairman should be independent". The Bahrain CG Code gives difference size of audit committee members as *per* Principle 3.1: "The board must establish a committee of audit of at least three members of [whom] should be a majority of independent including the chairman". It seems that the size of the committee of audit is also one of the important characteristics that govern the effectiveness of the committee of audit.

Consistent with the argument for an effective committee size, too small a committee size may mean that an insufficient number of directors are able to perform their work in the committee, and therefore the effectiveness of monitoring is reduced (Vafeas 2005). Evidence of committee sizes audit indicates that firms that have a larger committee of audit are more effective in monitoring management. Yang & Krishnan (2005) examine the relationship between quarterly management of earnings and size of the committee of audit in 896 American firms in the years 1996–2000. They found that the management of quarterly earnings is lower for firms that have a high number of committee of audit. This may indicate that the presence of adequate number of members of committee of audit increases the ability of the committee of audit in terms of monitoring the integrity of financial reports. In brief, the larger the size of a committee of audit, the more effective they are in the monitoring of financial reporting.

5.6 Independent committee of audit

Agency theory indicates that independence (of a director) is a fundamental quality that contributes to the effectiveness of a monitoring committee (Fama & Jensen 1983) and that the empirical evidence on the independence of the committee of audit is consistent with this proposal. Various studies suggest that the independent committee of audit are probably to be linked with the fraudulent financial reporting (Abbott et al. 2000, 2004) and more probably to be linked with lower earnings management (Klein 2002, Xie et al. 2003, Bedard et al. 2004, Davidson et al. 2005), and lower earnings restatement occurrence (Agrawal & Chadha 2005).

An independent committee of audit is expected to provide judgment and equitable assessment and to be able to monitor the management effectively. In brief, all of these suggest that independent committees of audit are linked with the higher quality of financial reporting and can be considered as effective monitors.

5.7 Audit committee expertise

According to the UK CG Code (2012), "The board should satisfy itself that at least one of the committee of audit members has financial experience" (C.3.1). DeZoort (1998) argues that the experience of the audit committee member in auditing and accounting is necessary to obtain a sufficient understanding of the oversight tasks. He proposes the following: "Audit and evaluation of internal control experience makes the difference in the members of the audit committee's performance on the internal control oversight task. It is important, the members of the committee of audit with the experience made internal control provisions more like those of experts (such as practicing auditors) without experience".

In other words, experimental evidence and regulatory concern indicate that the presence of knowledge and appropriate experience, particularly in the auditing and accounting, is likely to improve the audit committee's judgment and performance. The experimental evidence of archival studies also indicates that the financial expertise of the audit committee improves the ability of monitoring and results in an increase in the quality of financial reporting of firms. In general, all the assumption that supports the committee of audit with financial expertise has improved their effective monitoring function.

5.8 Committee of audit meeting

Various studies indicate that firms that have a larger number of meetings of the committee of audit less financial re-statement (Abbott et al. 2004), are less likely to be authorized for aggressive accounting and fraud (Abbott et al. 2000, Beasley & Petroni 2001) and are connected with a lower incidence of management of earnings (Xie et al. 2003). These studies indicate that committees of audit who meet often during the fiscal year related to effective monitoring. The more often they meet the more they increase their performance of their supervision duties. Therefore, the higher the number of meetings of the committee of audit, the more their monitoring function is improved.

6 SUMMARY

The previous section indicates that sole independence, more financial expertise, committee of audit with more members lead to a higher supervision function. Therefore, in line with the previous empirical evidence and agency theory proposition the hypotheses of this study show that these characteristics of committees of audit and boards are associated with a higher audit quality. With regard to the earnings management, this research views earnings management as opportunistic earnings.

Based on generic literature review, this research has developed a number of hypotheses to test. Taken together they have been designed to test the audit committee, board of directors, and auditor quality on constraining earnings management. The following are the hypotheses stated in a form that uses the non-audit services fees, audit fees, and the engagement of industry-specialist auditors as proxies for audit quality:

Hypothesis 1: The relationship between the non-audit services fees and size of the board shows a (positive) relationship.

Hypothesis 2: The relationship between the audit fees and size of the board shows a (negative) relationship.

Hypothesis 3: The relationship between the audit fees and independent board shows a (positive) relationship.

Hypothesis 4: The relationship between the engagement of industry-specialist auditors and the size of the board shows a (negative) relationship.

Hypothesis 5: The relationship between the engagement of industry-specialist auditors and an independent board shows a (positive) relationship.

Hypothesis 6: The relationship between the non-audit services fees and independent board shows a (negative) relationship.

Hypothesis 7: The relationship between the non-audit services fees and financial expertise of the board shows a (negative) relationship.

Hypothesis 8: The relationship between the audit fees and financial expertise of the board shows a (positive) relationship.

Hypothesis 9: The relationship between the audit fees and frequency of the meeting of the board shows a (positive) relationship.

Hypothesis 10: The relationship between the engagement of industry-specialist auditors and the financial expertise of the board shows a (positive) relationship.

Hypothesis 11: The relationship between the non-audit services fees and meeting frequency of the board shows a (negative) relationship.

Hypothesis 12: The relationship between the audit fees and size of the committee of audit shows a (positive) relationship.

Hypothesis 13: The relationship between the engagement of industry-specialist auditor and meeting frequency of the board shows a (positive) relationship.

Hypothesis 14: The relationship between the engagement of industry-specialist auditor and size of the audit committee shows a (positive) relationship.

Hypothesis 15: The relationship between the non-audit services fees and size of the audit committee shows a (negative) relationship.

Hypothesis 16: The relationship between the non-audit services fees and solely independent audit committee shows a (negative) relationship.

Hypothesis 17: The relationship between the audit fees and solely independent audit committee shows a (positive) relationship.

Hypothesis 18: The relationship between the audit fees and financial expertise of the audit committee shows a (positive) relationship.

Hypothesis 19: The relationship between the engagement of industry-specialist auditor and solely independent audit committee shows a (positive) relationship.

Hypothesis 20: The relationship between the engagement of industry-specialist auditor and financial expertise of the audit committee shows a (positive) relationship.

Hypothesis 21: The relationship between the non-audit services fees and financial expertise of the audit committee shows a (negative) relationship.

Hypothesis 22: The relationship between the non-audit services fees and frequency of the meeting of the audit committee shows a (negative) relationship.

Hypothesis 23: The relationship between the audit fees and frequency of a meeting of the audit committee shows a (positive) relationship.

Hypothesis 24: The relationship between the engagement of an industry-specialist auditor and frequency of the meeting of the audit committee shows a (positive) relationship.

These characteristics of the committee of audit and board are expected to constrain opportunistic earnings. In other words, this study tested the following hypotheses:

Hypothesis 25: The relationship between earnings management and an independent board shows a (negative) relationship.

Hypothesis 26: The relationship between earnings management and the board's size shows a (negative) relationship.

Hypothesis 27: The relationship between the earnings management and the board's meeting frequency shows a (negative) relationship.

Hypothesis 28: The relationship between the earnings management and board's financial expertise shows a (negative) relationship.

Hypothesis 29: The relationship between the earnings management and solely independent audit committee shows a (negative) relationship.

Hypothesis 30: The relationship between the earnings management and audit committee's size shows a (negative) relationship.

Hypothesis 31: The relationship between the earnings management and audit committee's meeting frequency shows a (negative) relationship.

Hypothesis 32: The relationship between the earnings management and audit committee's financial expertise shows a (negative) relationship.

Similarly, in line with the theoretical proposition and the review of evidence of differentiation of auditors' quality, this study showed that the effectiveness of audit services varies among auditors. In this paper, the higher auditor's quality is associated with the engagement of industry-specialist auditors, lower non-audit services fees, and higher audit fees. These expectations lead to the following hypotheses:

Hypothesis 33: The relationship between the management of earnings and non-audit services fees shows a (positive) relationship.

Hypothesis 34: The relationship between the management of earnings and industry-specialist auditor shows a (positive) relationship.

Hypothesis 35: The relationship between the management of earnings and audit fees shows a (negative) relationship.

Hypothesis 35: There is a (negative) relationship between the management of earnings and audit fees.

7 CONCLUSION

Corporate governance (CG) covers all the mechanisms that control the board of director and the audit committee, to lead them to improve the audit quality. In brief, good CG helps to prevent corporate scandals and fraud. In general, good CG is very important for firms suffering from poor reputations. It can make corporations more attractive for investors, customers, and other stakeholders.

REFERENCES

Abbott, L.J., Park, Y. & Parker, S. 2000. The effects of audit committee activity and independence on corporate fraud. *Managerial Finance* 26(11): 55–67.

Abbott, L.J., Parker, S. & Peters, G.F. 2004. Audit committee characteristics and restatements. *Auditing: A Journal of Practice & Theory* 23(1): 69–89.

Agrawal, A. & Chadha, S. 2005. Corporate governance and accounting scandals. *Journal of Law and Economics* 48(2): 371–406.

Albert, M. 1994. L'irruption du corporate governance. *Revue d'Economie Financiere* 31.

Barth, M.E., Elliott, J.A., Finn, M.W. 1999. Market rewards associated with pattern of increasing earnings. *Journal of Accounting Research* 37(2): 387–413.

Beasley, M.S. 1996. An empirical analysis of the relation between the board of director composition and financial statement fraud. *The Accounting Review* 71(4): 443–465.

Beasley, M.S. & Petroni, K.R. 2001. Board independence and audit-firm type. *Auditing: A Journal of Practice & Theory* 20(1): 97–114.

Beasley, M. & Salterio, S. 2001. The relationship between board characteristics and voluntary improvements in the capability of audit committees to monitor. *Contemporary Accounting Research* 18(4): 539–570.

Bedard, J., Chtourou, S.M. & Courteau, L. 2004. The effect of audit committee expertise, independence, and activity on aggressive earnings management. *Auditing: A Journal of Practice & Theory* 23(2): 13–35.

Bedard, J. & Gendron, Y. 2010. Strengthening the financial reporting system: Can audit committees deliver? *International Journal of Auditing* 14(2): 174–210.

Boo, E. & Sharma, D. 2008. Effect of regulatory oversight on the association between internal governance characteristics and audit fees. *Journal of Accounting and Finance* 48(1): 51–71.

Brennan, N. & McDermott, D. 2004. Alternative perspective on independence of director. *Corporate Governance: An International Review* 12(3): 325–336.

Burgstahler, D. & Dichev, I. 1997. Earnings management to avoid earnings decreases and losses. *Journal of Accounting and Economics* 24(1): 99–126.

Cadbury, A. 2000. The corporate governance agenda. *Corporate Governance* 8(1): 7–15.

Cahan, S.R. 1992. The effect of antitrust investigations on discretionary accruals: A refined test of the political-cost hypothesis. *The Accounting Review* 67(1): 77–95.

Carcello, J.V., Hermanson, D.R., Neal, T.L., & Riley Jr, R.A. 2002. Board characteristics and audit fees. *Contemporary Accounting Research* 19(3): 365–384.

Charreaux, G. 1997. "Vers une theorie du gouvernement des enterprises". In G. Charreaux (éd.), *Le gouvernement des enterprises*. Paris: Económica.

Chen, Y.M., Moroney, R. & Houghton, K. 2005. Audit committee composition and the use of an industry specialist audit firm. *Accounting and Finance* 45(2): 217–239.

Chtourou, S.M., Bedard, J. & Courteau, L. 2001a. Corporate Governance and Earnings Management. Available at http://ssrn.com/abstract=275053.

Chtourou, S.M., Bédard, J. & Courteau, L. 2001b. *Corporate Governance and Earnings Management*. Université Laval, Canada.

Clarkson, P.M. & Simunic, D.A. 1994. The association between audit quality, retained ownership, and firm-specific risk in US vs. Canadian IPO markets. *Journal of Accounting and Economics* 17: 207–228.

Cohen, J., Krishnamoorthy, G. & Wright, A. 2002. Corporate governance and the audit process. *Contemporary Accounting Research* 19(4): 573–594.

Collier, P. & Gregory, A. 1996. Audit committee effectiveness and the audit fees. *The European Accounting Review* 5(2): 177–198.

Conger, J., Finegold, D. & Lawler III, E. 1998. Appraising boardroom performance. *Harvard Business Review* 76: 136–148.

Davidson, R., Goodwin-Stewart, J. & Kent, P. 2005. Internal governance structures and earnings management. *Accounting and Finance* 45(2): 241–267.

DeAngelo, L. 1981. Auditor size and audit quality. *Journal of Accounting and Economics* 3(3): 183–199.

DeFond, M.L., Hann, R.N. & Hu, X. 2005. Does the market value financial expertise on audit committees of boards of directors? *Journal of Accounting Research* 43(2): 153–193.

DeFond, M.L. & Jiambalvo, J. 1994. Debt covenant violation and manipulation of accruals. *Journal of Accounting and Economics* 17(1–2): 145–176.

DeZoort, F.T. 1998. An investigation of audit committees' oversight responsibilities. *ABACUS* September: 208–227.

Fama, E.F. 1980. Agency problems and the theory of the firm. *Journal of Political Economy* 88(2): 288–307.

Fama, E.F. & Jensen, M.C. 1983. Separation of ownership and control. *Journal of Law and Economics* 26: 301–325.

Gaver, J.J., Gaver, K.M. & Austin, J.R. 1995. Additional evidence on bonus plans and income management. *Journal of Accounting and Economics* 19: 3–28.

Habbash, M., Salama, A., Dixon, R. & Hussainey, K. 2010. The effects of nonexecutive directors' commitment, chairman independent and ownership structure on earnings management. *Journal of Applied Accounting Research* 13ii.

Han, J.C.Y. & Wang, S.-W. 1998. Political costs and earnings management of oil companies during the 1990 Persian Gulf crisis. *The Accounting Review* 73(1): 103–117.

Hasan, S.U. & Ahmed, A. 2012. Corporate governance, earnings management and financial performance: A case of Nigerian manufacturing firms. *American International Journal of Contemporary Research* 2(7).

Healy, P.M. 1985. The impact of bonus schemes on the selection of accounting principles. *Journal of Accounting and Economics* 7: 85–107.

Healy, P.M. & Wahlen, J.M. 1999. A review of the earnings management literature and its implications for standard setting. *Accounting Horizons* 13(4): 365–383.

Holthausen, R.W., Larkers, D.F. & Sloan, R.G. 1995. Annual bonus schemes and the manipulation of earnings. *Journal of Accounting and Economics* 19: 29–74.

Jensen, M.C. 1993. The modern industrial revolution, exit, and the failure of internal control systems. *The Journal of Finance* 48(3): 831–880.

Jones, J. 1991. Earnings management during import relief investigations. *Journal of Accounting Research* 29(2): 193–228.

Kalbers, L.P. & Fogarty, T.J. 1993. Audit committee effectiveness: An empirical investigation of the contribution of power. *Auditing: A Journal of Practice & Theory* 12(1): 24–48.

Klein, P. 2002. Audit committee, board of director characteristics, and earnings management. *Journal of Accounting and Economics* 33(3): 375–400.

Lin, J.W. & Hwang, M.L. 2010. Audit quality, corporate governance and earnings management: A meta-analysis. *International Journal of Auditing* 14: 57–77.

Lipton, M. & Lorsch, J.W. 1992. A modest proposal for improved corporate governance. *The Business Lawyer* 48: 59–77.

Menon, K. & Williams, J. 1994. The use of audit committees for monitoring. *Journal of Accounting and Public Policy* 13(2): 121–139.

Peasnell, K.V., Pope, P.F. & Young, S. 2005. Board monitoring and earnings management: Do outside directors influence abnormal accruals? *Journal of Business Finance & Accounting* 32(7–8): 1311–1346.

Ronen, J. & Yaari, V. 2008. *Earnings Management: Emerging Insights in Theory, Practice and Research*. Springer: New York.

Ruddock, C., Taylor, S.J. & and Taylor, S.L. 2006. Non-audit services and earnings conservatism: Is auditor independence impaired? *Contemporary Accounting Research* 23(3): 701–746.

Schipper, K. 1989. Commentary on earnings management. *Accounting Horizons* 3(4): 91–102.

Shleifer, A. & Vishny, R.W. 1997. A survey of corporate governance. *Journal of Finance* 52.

Sweeney, A. 1994. Debt covenant violations and managers' accounting responses. *Journal of Accounting and Economics* 17(3): 281–308.

Tierney 2006. *Governance and the Public Good*. New York: State University of New York Press.

UK Corporate Governance Code 2010. London: Financial Reporting Council.

UK Corporate Governance Code 2012. London: Financial Reporting Council.

Vafeas, N. 1999. Board meeting frequency and firm performance. *Journal of Financial Economics* 53: 113–142.

Vafeas, N. 2005. Audit committees, boards, and the quality of reported earnings. *Contemporary Accounting Research* 22(4): 1093–1122.

Wallace, W.A. 1980. *The Economic Role of the Audit in Free and Regulated Markets*. New York: Touche Ross.

Watts, R.L. & Zimmerman, J.L. 1986. *Positive Accounting Theory*. Englewood Cliffs, NJ: Prentice-Hall.

Xie, B., Davidson, W.N. & DaDalt P.J. 2003. Earnings management and corporate governance: The role of the board and the audit committee. *Journal of Corporate Finance* 9(3): 295–316.

Yang, J.S. & Krishnan, J. 2005. Audit committee and quarterly earnings management. *International Journal of Auditing* 9: 201–219.

Yermack, D. 1996. Higher market valuation of companies with a small board of directors. *Journal of Financial Economics* 40: 185–202.

Review of energy management policies in healthcare buildings

Salman Shehab
Ahlia University, Kingdom of Bahrain

ABSTRACT: Global energy issues relating to industry, transportation and building are reviewed in this paper. Consumption by buildings, both residential and commercial, is considered. Hospitals and healthcare facilities as major energy intensive and energy demanding buildings are selected to be the research area. The purpose of this paper is to define a gap and generate a problem statement related to the research subject by conducting a systematic literature review of existing energy management policies in hospitals and healthcare facilities which are seen to use the terms 'energy efficiency' and 'energy conservation' interchangeably. It is also found that these policies are technology driven with disregard to energy conservation (human preferences, behaviour and motivational changes). The paper highlights the need to develop new policy that can be used as a strategic decision-making and risk-assessment tool in handling energy issues to achieve sustainable economic and environmental goals.

1 INTRODUCTION

1.1 *Overview*

The world's energy is consumed by three major sectors, namely industry, transportation and buildings. Although these three sectors share the same environmental impacts, each sector has its own characteristics connected to the trend of energy consumption that leads to this classification.

The rapidly growing world energy use has raised concerns over supply difficulties of energy resources and heavy environmental impacts (ozone layer depletion, global warming, climate change, etc.). The global contribution from buildings towards energy consumption, both residential and commercial, has steadily increased reaching figures between 20% and 40% in developed countries, and has exceeded the other major sectors, industrial and transportation. Growth in population, increasing demand for building services and comfort levels, together with the rise in time spent inside buildings, assure that the upward trend in energy demand will continue. For this reason, energy efficiency in buildings is today a prime objective for energy policy at regional, national and international levels (Perez-Lombard et al. 2008) and changing behaviour patterns from one side and technology from the other are key issues for public energy policy (Oikonomou et al. 2009).

Buildings, both residential and commercial, can be further classified based on the types of activity that determine the characteristics of its operation, energy consumption and consequently the suitable energy management policy. The sub-classification would include shopping malls, banks, office buildings, universities, schools, hospitals, health centres, etc.

1.2 *Background of the research topic*

Hospitals and other healthcare facilities are considered as major energy intensive and energy demanding public buildings due to, but not limited to, 24-hour operation every day of the week (medical equipment's power, operating conditions requirements, tight ventilation and air quality necessities). A study in the American healthcare sector found hospitals are responsible for 9% of the total energy consumption (Perez-Lombard et al. 2008) and 7% of carbon dioxide (CO_2) emission (Chung et al. 2009). The National Health Service (NHS) in England is responsible for 6% of the total energy consumption (Perez-Lombard et al. 2008) and 30% of all public CO_2 emissions (Gatenby 2011).

This demand is expected to grow significantly in the forthcoming years due to the construction of new facilities, the expansion of buildings and the renovation of facilities. This growth is to meet the increase in healthcare demands due to demographical growth, demographical shifting and new healthcare regulations and legislations. The existing energy management policies related to healthcare need to be reviewed and developed to meet the challenges related to this growth.

The aim of this paper is to define a gap and generate a problem statement related to the research subject by conducting a systematic literature review of existing energy management policies in hospitals and healthcare facilities.

2 METHODOLOGY

This research is initiated by conducting a systematic review, as described by Tranfield et al. (2003), to cover preceding efforts and works related to the research topic. The main area of research is the energy policy in buildings, particularly hospitals and healthcare facilities. The keywords and research terms used are a logical combination of 'energy, policy, management, saving, efficiency, conservation, hospitals, healthcare, buildings and facilities'. The search includes published journals, unpublished journals, conference proceeding, reliable Internet resources, etc. The search was limited to studies carried out between 2004 and 2014.

3 LITERATURE REVIEW

The purpose of this section is to cover the following research related aspects: energy management schemes in hospitals and healthcare buildings, energy saving measures, energy efficiency drivers and barriers, energy efficiency services, energy conservation and the rebound effect.

3.1 *Energy management schemes in hospitals and healthcare buildings*

Energy management schemes in hospitals and healthcare buildings are part of 'sustainability plan'. They cover a diversified range of activities related to energy consumption. They also target a number of sustainable economic and environmental goals such as reduction of energy usage, reduction of related greenhouse gases and carbon dioxide emissions to reduce a building's carbon footprint, and the reduction of energy costs to reach the desired operating budget. These schemes are very important, as it is the research subject covered by the energy policy.

3.2 *Energy saving measures*

Energy saving addresses the reduction of final energy consumption through energy efficiency improvement or behavioural change [energy conservation] (Oikonomou et al. 2009). The terms "energy efficiency" and "energy conservation" have often been used interchangeably in policy discussion but they do have very different meanings (Herring 2006). Energy efficiency is a term widely used, often with different meanings in public policy making. A clear distinction between energy efficiency and energy conservation is that the former refers to the adoption of a specific technology that reduces overall energy consumption without changing the relevant behaviour, while the latter implies merely a change in consumers' behaviour (Oikonomou et al. 2009).

In this regard, the authors would like to emphasize the structure of an energy saving process that comprises two dissimilar measures, energy efficiency and energy conservation. Using these two terms interchangeably in policy discussion and in academic and practical researches is misleading and may lead to misjudgment of the nature of each term and subsequently misjudgments of the suitable management process. Using the term energy efficiency with different meanings in public policy making may lead to the same results while equating energy efficiency to energy saving in the energy literature may lead to disregard of energy conservation and subsequently disregarding its impact on the energy saving process. As 'energy efficiency' and 'energy conservation' form the energy saving process that is at the heart of this research, it is advisable to study it in more detail.

3.2.1 *Energy efficiency*

Energy efficiency, as noted by Oikonomou et al. (2009), concerns the technical ratio between the quantity of the primary or final energy consumed and the maximum quantity of energy services obtainable (heating, lighting, cooling, mobility, etc.). It is simply the ratio of energy services out to energy in (Schipper et al. 1997). Various studies indicate that increased energy efficiency can bridge the gap between growing demand and reduced energy supply without adversely affecting the quality of service. This may not happen, however, because there is a gap between the theoretical opportunities for cost-effective energy efficiency investment and the level that can be achieved practically (Sudhakara 2013). This is a very important factor affecting the proposed framework. It can be measured before and after implementing of specific preference technology.

3.2.2 *Energy conservation*

The concept of energy conservation refers to the reduction of energy consumption associated with a frugal lifestyle that includes a form of regulation (speed limitation, reduced domestic heating, and so on) or spontaneous change in consumer preferences resulting in behaviour changes (Oikonomou et al. 2009). This concept often implies more moral aspects of behaviour rather than a strictly economic one. It can be enhanced *via* changes in the context (including regulations and energy price increases) and changes in motivation of people (including environmental concerns and a feeling of moral obligation to reduce energy consumption); it is reducing energy consumption through lower quality of energy services (Schipper & Haas 1997). This is another important factor affecting the proposed framework. It can be measured before and after implementing regulatory changes or limitations of energy sources. It is also can represent, in a numerical form, the change in behaviour and motivation of people.

3.3 *Energy efficiency drivers and barriers*

While energy conservation measures utilize low quality of energy services such as thermostat- or switch-setting to reduce energy consumption, energy efficiency measures need heavy investment in technology to achieve this goal. Energy efficiency technologies normally face some barriers that can be considered as obstacles to investment. On the other hand many drivers can be considered as the factors that promote investment in energy efficiency.

3.3.1 *Energy efficiency drivers*

Hospitals and healthcare organizations are moving toward energy efficiency investment due to different significant motivating factors (drivers). These factors as illustrated by Smith (2011) are to reduce long-term cost, create a safer working environment for staff, create a more comfortable environment for patients, promote the facility and enhance brand image to attract staff, patients and donors, respond to existing or anticipated legislations, reduce carbon footprint and the green house effect (sustainability), meet community expectations toward environmental commitments, and finally to secure government and utility grants and incentives. Energy efficiency motivators, especially those related to regulatory or legislative bodies, are important drivers to energy management programmes.

3.3.2 *Energy efficiency barriers*

Significant growth in the construction activities of hospitals and healthcare buildings will lead to continued increase of energy demand. Expansion of these facilities' operations will lead to continuous increase of consumption. Herring (2006) believed that energy efficiency will not necessarily lead to a reduction in energy use and hence a reduced CO_2 emission. It will mean, however, a safe move toward a fossil-free energy future. Energy consumption's annual increment rate in addition to other internal and external drivers are determining the trend of future energy demand and consumption.

Hospitals and healthcare buildings are investing in energy efficiency to compensate for the increase of energy demand. Hospitals and healthcare organizations face several barriers preventing them from investing to their desired extent to reach their full energy saving potential. The largest of these barriers is finance. Insufficient payback, or uncertainty that project return on investment would be realized, is another challenge (Smith 2011). Lack of investment and project financing can delay the start and affect the future progress of energy saving programmes.

3.4 *Energy efficiency investment*

To improve energy efficiency in hospitals and healthcare facilities many buildings' electromechanical services can be optimized by incorporating energy efficiency technologies either at the initial stage or by retrofitting these systems. A study by Smith (2011) shows that improving heating, ventilation and air conditioning (HVAC) systems and lighting system accounts for approximately 60% of all energy used in traditional buildings. These top two improvements alone can help healthcare facilities move a long way toward energy efficiency. Moreover, embracing new technologies that continue to grow in popularity and align with clean energy solutions such as solar water heating can add to the saving while optimizing other services can maximize energy saving achievement.

The focus on HVAC systems is due to the significant growth in energy used which reached 53% of building consumption in the USA, 68% in the EU (as a whole) and 62% in the UK. Lighting systems are the second area of focus, where energy consumption is found to be 30% in the USA, 18% in the EU, and 16% in the UK of building consumption. Domestic hot water is found to be another important area of energy consumption with 17% of consumption in the USA, 14% in the EU, and 22% in the UK of building consumption (Perez-Lombard et al. 2008).

3.5 *Energy conservation and the rebound effect*

Another important factor affecting energy efficiency in hospitals and healthcare buildings is that some of the saving from efficiency improvements will be taken in the form of higher energy consumption, the so-called take-back or rebound effect. There is, however, intense dispute over its magnitude and it is strongly suspected that it is much less than 100%, perhaps of the order 10–20% (Herring 2006). A number of recent empirical studies have begun to establish a very strong body of evidence that point to rebound effects being relatively small, in the order of 5–10% (Saunders 1992). The rebound is not high enough to mitigate the importance of energy efficiency as a way of reducing carbon emissions. Climate policies that rely only on energy efficiency technologies may, however, need reinforcement by market instruments such as fuel taxes and other incentive mechanisms. Without such reinforcements, a significant portion of the technologically achievable carbon and energy savings could be lost to the rebound (Greening et al. 2000).

Although the rebound effect has a counter-impact on energy-efficiency improvement (adoption of technologies), it is mainly linked, as a negative behaviour, to the energy conservation that requires incentive mechanism or motivation to balance it. The rebound effect is an important factor and disregarding it in policy development may lead to an unfavourable effect on energy saving process.

The literature related to researches on rebound effects in buildings' energy shows a focus on the phenomenon in household and residential energy consumption (Galvin 2014, Yu et al. 2013, Lin & Liu 2013a, b, Chitnis et al. 2013) and on energy consumption in the transportation sector

(Lin & Liu 2013a, b, Wang et al. 2012; Evans & Schafer 2013) but no dedicated studies found directly related to healthcare buildings. As healthcare buildings have the same services and household equipment, the data related to these items can be used where applicable to develop the healthcare energy policy.

4 DISCUSSION AND ANALYSIS

As discussed earlier, the energy saving process in buildings, particularly hospitals and other healthcare buildings, comprises of two dissimilar measures, that is, energy efficiency and energy conservation. Using these two terms interchangeably in energy policy discussion and in academic and practical researches is misleading and may cause misjudgments of the nature of each term and subsequently an error in selecting the suitable mitigation solution. Using the term energy efficiency with different meanings in public policy making may lead to the same unfavourable results.

Working toward energy management or energy saving is supported by national, regional and international organizations that produce white papers, case studies, reports and roadmap plans. It is also highly supported by consultancy firms offering consultancy services and by the industrial sector that provides technical solutions supported by successful experiences and case studies. These firms make the energy saving process highly driven by technology and ignoring the influence of other measures. Equating energy efficiency to energy saving in the energy literature may lead to disregard of energy conservation measures, subsequently disregarding its impact on the energy saving process.

As the research emphasis is on consideration of energy conservation side-by-side with energy efficiency in any energy policy development, the need to consider a counter-effect phenomenon called 'pay back or rebound effect' is raised, especially in similar studies dedicated to hospital and other healthcare buildings.

5 ALTERNATIVE POLICY

The existing energy management policies, implemented in hospitals and healthcare buildings, give attention to some important parameters such as 'energy conservation' and 'rebound effect'. These policies need to be reviewed in order to develop a new policy. The new developed policy must take into account, in addition to the adoption of technology, human preferences, human behaviour and motivational changes related to the use of energy in hospitals and other healthcare facilities.

The policy developed can be used as a strategic decision-making and risk-assessment tool in handling energy issues to achieve sustainable economic and environmental goals. The targeted sustainable economic goal is the reduction of energy costs that contribute directly to reducing the hospitals' and other healthcare facilities' operating budgets. A sustainable environmental goal is the reduction of energy consumption, which leads to a reduction of related greenhouse gases and carbon dioxide emissions, and subsequently reduction of a building's carbon footprint. The research is also seeking academic contributions in developing hospitals' and healthcare energy management programmes by providing scientific persuasive modelling.

6 CONCLUSIONS

Hospitals and healthcare buildings are increasingly going toward energy management schemes as part of their sustainability plan and to meet certain circumstances. The majority of researches concerning energy management programmes are led by national, regional and international organizations, industry and consultancy firms. It is focussing on technical energy saving measures and disregarding some other parameters that is influencing the outcomes of these programmes such as human preferences, human behaviour and motivational changes related to the use of energy in hospitals and other healthcare facilities. The energy management policies in hospitals and healthcare buildings need to be reviewed in order to develop a new policy. The new developed policy will take into account all factors influencing the energy management process in hospitals and healthcare buildings. The developed policy can be used as a strategic decision-making and risk-assessment tool in handling energy issues to achieve sustainable economic and environmental goals.

Future activities pertaining to this research study comprise, developing the conceptual framework, developing a model representing different parameter relations and selecting a model analysis method. This method can be used in analysing energy parameters in addition to solving current and future energy-related issues.

REFERENCES

Chitnis, M., Sorrell, S., Druckman, A., Firth, S.K. & Jackson, T. 2013. Turning lights into flights: Estimating direct and indirect rebound effects for UK households. *Energy Policy* 55: 234–250.
Chung J.W. & Meltzer D.O. 2009. Estimate of the carbon footprint of the US healthcare sector. *J. Am. Med. Ass.* 302.
Evans, A. & Schafer, A. 2013. The rebound effect in the aviation sector. *Energy Economy* 36: 158–165.
Galvin, R. 2014. The 'rebound effect' more useful for performance evaluation of thermal retrofits of existing homes: Defining the energy savings deficit and the energy performance gap. *Energy Build* 69: 515–524.
Gatenby P.A. 2011. Modeling carbon footprint of reflux controls. *Int. J. Surg.* 9: 72.
Greening, L.A., Greene, D.L. & Difiglio, C. 2000. Energy efficiency and consumption – the rebound effect – a survey. *Energy Policy* 28(6–7): 389–401.
Herring, H. 2006. Energy efficiency: A critical view. *Energy* 31: 10–20.
Lin, B. & Liu, X. 2013a. Electricity tariff reform and rebound effect of residential electricity consumption in China. *Energy* 59: 240–247.
Lin, B. & Liu, X. 2013b. Reform of refined oil product pricing mechanism and energy rebound effect for passenger transportation in China. *Energy Policy* 57: 329–337.
Oikonomou, V., Becchis, F., Steg, L. & Russolillo, D. 2009. Energy saving and energy efficiency concepts for policy making. *Energy Policy* 37(11): 4787–4796.
Perez-Lombard, L., Ortiz, J. & Pout, C. 2008. A review on buildings energy consumption information. *Energy & Building* 40(3): 394–398.
Saunders, H. 1992. The Khazoom-Brooks postulate and neoclassical growth. *Energy Journal* 13: 131–148.
Schipper L. & Haas R. 1997. Cross-country comparisons of indicators of energy use, energy efficiency and CO_2 emissions. *Energy Policy* (special issue) 25: 7–9.
Smith, R. 2011. Energy management opportunities and challenges for healthcare industry. *Healthcare Financial Management Association* 65(4): 98–102.
Sudhakara R.B. 2013. Barriers and drivers to energy efficiency: A new taxonomical approach. *Energy Conversion and Management* 47: 403–416.
Tranfield, D., Denyer, D. & Smart 2003. Towards a methodology for developing evidence-informed management knowledge by means of systematic review. *British Journal of Management* 14: 207–222.
Wang, H., Zhou, D.Q., Zhou, P. & Zha, D.L. 2012. Direct rebound effect for passenger transport: Empirical evidence from Hong Kong. *Appl. Energy* 92: 162–167.
Yu, B., Zhang, J. & Fujiwara, A. 2013. Evaluating the direct and indirect rebound effects in household energy consumption behavior: A case study of Beijing. *Energy Policy* 57: 441–453.

A data mining approach for investigating students' completion rates

Subhashini Bhaskaran
Ahlia University, Kingdom of Bahrain

Kevin Lu
Brunel University London, UK

Mansoor Al Aali
Ahlia University, Kingdom of Bahrain

ABSTRACT: One of the major challenges faced by higher education institutions is to enhance the quality of decisions made from knowledge derived from rapidly growing educational data. Data mining techniques are investigative tools that are used to extract significant unknown information from large data sets. This paper proposes to discover the most appropriate data mining technique(s) to investigate the relationship between prior learning, temporal sequence of courses and student performance attributes namely GPA and time-to-degree (number of semesters taken towards graduation) and later the correlation between GPA and time-to-degree. Once the relationships are established, it is proposed to find the optimized sequence of courses taken by successful students from similar prior learning backgrounds that would facilitate current/future students to graduate on time with high scores. More specifically this paper highlights this research gap from the literature review which will be further analysed by the authors using data mining.

1 INTRODUCTION

Modern day higher education institutions (HEIs) are faced with serious challenges which threaten their very existence. For instance, factors such as globalization of education, severe competition amongst institutions and increasing demands by students are forcing HEIs to reform in order to counter the challenges, as stated at the UNESCO forum on HEIs (WEF 2000). Although many researchers argue that there is a vital need to address this situation through in-depth research, the literature shows that implementable outcomes through research are hard to find. Even the few that have been touted to be successful are afflicted by serious limitations in terms of context, methodology and generalizability. One such direction of research that has attracted researchers and practitioners in the HEIs alike is the use of data mining techniques leading to better capability of HEIs to face the challenges successfully.

1.1 Data mining

Data mining is defined as 'the process of finding interesting hidden knowledge from huge volumes of data using algorithms'. Data mining belongs to the family of machine learning, statistics and artificial intelligence. Data mining has created a lot of interest in the information industry because of the existence of voluminous data and increasing demand for converting these data into useful knowledge for use in the analysis of market, business and decision support (Fayyad et al. 1996b).

While data mining as an area of research has attracted a large number of researchers (Romero & Ventura 2007, Baker & Yacef 2009, Vialardi et al. 2011), much still needs to be done to gain a complete understanding of the benefits of applying data mining to a variety of fields including higher education. For instance Huebner (2013) argues that data mining techniques need to be applied to the field of higher education in order to facilitate an understanding of many factors that influence student performance, decision making and improving efficiency. Thus it can be seen that a major area that needs further study in the field of higher education is the application of data mining to enable institutions to improve their performance. This research is one such effort to find a method by which data mining techniques could be used to understand how student performance could be enhanced with regard to time to degree and optimized sequence of study of course from voluminous educational data.

1.2 *Problem statement*

It can be seen from the existing literature that there is a need to investigate more dominating factors that contribute to student performance, like the sequence of courses, prior learning, time to degree, GPA and the inter-relationship between them using data mining which could produce promising results and insight into student performance. Furthermore there is paucity of study that has analysed the relationship between prior learning, sequence of courses with time to degree and student performance (GPA) together in a single research or study. This knowledge of the best or optimized sequence of courses and better prior learning understanding could possibly lead to on-time to degree completion and higher GPA and will help students and educators in effectively and efficiently completing the programmes.

1.3 *Aim and research questions*

Having the above problem statement in mind, the aim of the research is as follows

- To develop or identify most appropriate technique(s)/function(s) that could address the purpose of gaining a greater understanding of factors that could lead to better productivity and efficiency of the students in HEI.

The above aim can be accomplished when the following questions are answered

- How can data mining be used to discover patterns of sequence of courses that students groups from similar prior learning background should take
    - to get a high GPA, and
    - finish the degree on time?

1.4 *Methodology*

The steps taken will be to

- investigate the current data mining techniques to develop an enhanced data mining technique that could be applied to factors relating to student performance in HEI,
- apply the newly developed model for data mining and knowledge discovery,
- test the developed technique to address the factors and the research questions,
- provide conclusions and recommendations.

1.5 *Expected findings*

This investigation is expected to produce better insight into the dominating factors that contribute to student performance. It will help in the better understanding of whether factors like sequence of courses and prior learning affect student performance attributes like student GPA and time to degree. Also the inter-relationship amongst the factors will help in understanding if any of these factors affect the other factors. For example, if prior learning affects sequence of courses or not and if found to be affecting then the curriculum specialists can design the programme plan based on the prior learning of the students.

1.6 *Theoretical contribution*

The paper makes the following contributions to knowledge

- whether factors like course sequence and prior learning affect student GPA and time to degree and the inter-relationships between them,
- The process or steps or methodology of finding such knowledge using data mining and development of model are revealed.

1.7 *Practical implications*

The research reports the following implications arising from the research

- This knowledge can help the academicians or management of HEI to plan the curriculum and admit students accordingly with the right prior learning background so that the students could finish the programme on time with high GPA which would yield better placements.
- The HEIs could revisit their admission procedure(s) to admit students with appropriate prior learning or create orientation courses for students based on prior learning background to make them equipped for the programme so they can finish them on time with high GPA.
- Because of on time completion of students better will be the graduation rates which will lead to better admission rates and better profitability for both the students and HEI.

2 CURRENT STATUS OF HIGHER EDUCATION INSTITUTIONS

Higher Educational Institutions (HEIs) are undergoing major transformations (UNESCO forum on HEIs). Various challenges like the rise of the knowledge economy, economic globalization, the increasing demand for quality education, demands of the job market and the growing number of HEIs contribute to these transformations (Kowalkiewicz 2007, Hénard 2008). The challenges faced by HEIs need to be successfully overcome if they want to survive in a highly competitive world. It won't be an exaggeration to state that successful negotiation of the challenges by HEIs can only be possible if the HEIs enhance the quality of their performance continuously based on regular measurement of their performance and transform their present situation to a more viable one.

The literature on HEIs (Martens & Prosser 1998, Becket & Brookes 2008) and lectures at the World Education Forum in Dakar in 2000 show that measuring the quality performance of the HEIs objectively is a major challenge. In fact measuring the quality of performance has been argued by many researchers (Becket & Brookes 2006, 2008, Eriksen 1995, Oldfield & Baron 1998, Cheng & Tam 1997) as a contentious issue in the context of HEIs. For instance McKay & Kember (1999), Ramsden (1992), Martens & Prosser (1998) and Yorke (1999) argue that measuring the quality of teaching is a major concern of HEIs across the world. Similarly, measuring the extent of excellence achieved through the performance of the HEIs is considered to be another challenge.

Researchers have been seriously involved in finding a way forward to tackle these challenges. For example the OECD report on quality in teaching (OECD 1994) has argued that certain aspects of teaching could be objectively assessed and evaluated to understand the quality of teaching although the research outcomes are not conclusive. Similarly, many researchers have used the EFQM model to assess the excellence achieved by HEIs although the majority of the studies have produced subjective results and therefore need further study.

The above discussion clearly points out that measuring the quality of performance of HEIs is a major concern of many researchers and is an important topic that requires investigation to help HEIs to transform themselves into institutions that perform better. The literature in the field of HEIs indicates that a number of performance aspects are under investigation by researchers which include enhancing student performance, quality of teaching, quality of learning, curricula issues and student satisfaction. The need to investigate those aspects arises from the fact that the literature does not cover all types of HEIs or all the students studying in different systems of education (Dew 2009).

Amongst the various aspects that influence the HEIs in measuring their quality of performance is the methodology aspect as there is a lack of unanimity amongst researchers (Owlia & Aspinwall 1996, Borahan & Ziarati 2002, Badri & Abdulla 2004, Mizikaci 2006) on which one of the methods highlighted in the contemporary literature is the most suitable. For instance many researchers have suggested that self-evaluation techniques can be used by HEIs to measure the quality of performance although it has been criticized by many as merely subjective. Other researchers have resorted to empirical studies linking factors that could be measured objectively to the quality of performance of the HEIs although the results of such research are not generalizable because the results are particular to specific contexts chosen by the researchers. Despite such serious limitations, there is a growing interest shown by the research community to adopt such techniques as data mining to understand and measure the quality of performance of HEIs although this area is relatively new. The limited success achieved by researchers in applying data mining techniques to understand the quality of performance of HEIs is sufficiently encouraging to conduct further research in this area.

Research in data mining techniques has been of interest amongst researchers and literature shows that in the field of HEIs researchers have been contributing significantly by studying and applying data mining techniques to business aspects (Castellano & Martínez 2008, Cortez & Silva 2008, Al-Radaideh et al. 2006, Waiyamai 2003, Luan 2001, 2002a, b).

3 LITERATURE REVIEW

3.1 Student performance factors

In the field of higher education, both researchers and practitioners are in agreement on the need to enhance both student performance and institutional performance by investigating many relevant factors as highlighted in Section 2. Current knowledge in those factors that affect student performance and institutional performance has been argued to be inadequate leading to the need to gain more knowledge about those factors. For instance, from the extant literature it can be found that some researchers (Kotsiantis et al. 2004, Ramaswami & Bhaskaran 2010) have found key demographic variables and assignment marks to be significant, while Woodman (2001) found that the mark for the first assignment was the significant factor that affects student performance. Simpson (2006) found course level and credit rating to be significant while Herrera (2006) found academic level to be important. Then researchers like Vandamme et al. (2007), Cortez & Silva (2008), Lykourentzou et al. (2009) and Ramaswami & Bhaskaran (2010) found previous education or prior learning to be a predominating factor in student performance. Furthermore, Quadri & Kalyankar (2007) found low income level of students and ethnicity to be significant whereas Kumar & Vijayalakshmi (2011) found internal assessment to contribute to student success. Bharadwaj & Pal (2011) found previous semester marks to be a major factor while Yu et al. (2007) found cumulated earned hours as important.

Learning style is significant and Siraj & Abdoulha (2009) found faculty and nationality to be important as did Dekker et al. (2009) about the secondary school science mark. Communication, learning facilities, guidance and family stress are significant while Yadav et al. (2011) found graduate stream to be important. Lykourentzou et al. (2009) found gender to be a principal determinant whereas Yu et al. (2007) found gender to be insignificant in predicting student performance.

From the above discussion it can be seen that even though much research has been done to analyse student performance there has been no consensus on the factors that could affect it. Current efforts in the field of higher education have not been able to address many other aspects and contexts, for instance understanding the enrolment characteristics of students at the entry point as a function of their length of study in the university. In fact many researchers (for example, Romero & Ventura 2007) have highlighted the need to extend data mining techniques to many different aspects of education in the HEI sector as institutions are not able to make strategic or critical decisions in regard to improving the efficiency or productivity of the institution.

According to Huebner (2013) and Osmanbegović & Suljić (2012) further research is needed in the HEIs in order to identify more factors affecting student performance using such tools as data mining, the result of which could be very useful to HEIs in decision making. Thus there is a need to examine how data mining techniques could be enhanced and a model developed to address some of the vital issues concerning the HEIs and the students.

These arguments clearly indicate that there are significant areas in the literature pertaining to the HEIs in general and data mining in particular that need further investigation. This paper investigates this area in the literature and through this research it is expected that new knowledge will be added to the current literature.

3.2 *Literature on student performance analysis*

Table 1 shows data mining techniques and algorithms used by authors of research papers on analysing student performance. From this table it can be seen that different techniques were used for the same functionality and the same technique for different functionality. It can be seen that there is no consensus in the techniques and algorithms that were used.

Table 2 lists the classification techniques and algorithms used in student performance investigations. From this table it can be seen that classification technique mainly algorithms like CART, naïve Bayes and CHAID tree were preferred by researchers working in educational data mining, specifically in student performance investigations.

It can be seen from extant literature that many algorithms in classification techniques like artificial neural network algorithms such as SLP, COHONEN, MNP are rarely used. Additionally, algorithms from associated techniques, namely RELIM, FP-GROWTH and from clustering

Table 1. Taxonomy of data mining techniques used for student performance analysis.

Educational functionality	Data mining technique	Data mining algorithm	Researchers
Student performance analysis (prediction and factors)	Classification	ID3 decision tree algorithm	Osmanbegović & Suljić 2012, Yadav et al. 2011.
		Bayes classification	Bhardwaj & Pal (2012)
		C4.5 decision tree	Kumar & Vijayalakshmi (2011)
		Rule induction and naïve Bayesian classifier	Tair & El-Halees (2012)
		CHAID Tree	Ramaswami & Bhaskaran (2010)
		CHAID and CART trees	Ramaswami & Bhaskaran (2010)
		Simple regression	Anwar & Ahmad 2011, 2013, Tair & El-Halees 2012.
		A priori	Tair & El-Halees (2012)
	Regression technique Association techniques Clustering technique	K-means	
Student persistence and drop outs	Classification	Neural network, support vector machines	Lykourentzou et al. 2009
	Regression techniques		Woodman 2001, Herrera 2006, Quadri & Kalyankar 2007
Recommender systems	Classification	C4.5, KNN and naïve Bayes	Vialardi et al. 2011
Student placement	Classification	Decision tree and naïve Bayes	Pal & Pal 2013

Table 2. Classification techniques and algorithms used in student performance investigations.

Educational functionality	Best algorithm	Compared algorithms	Researchers
Recommender system	C4.5	C4.5, KNN and naïve Bayes	Vialardi et al. (2011)
Student performance analysis	CHAID tree	Simple regression	Ramaswami & Bhaskaran (2010)
	CART Decision tree Naïve Bayes classification CART	CHAID Bayesian network algorithm Decision tree and neural networks ID3 and C4.5	Osmanbegović & Suljić (2012)
	ID3	C4.5 and bagging technique	Pal & Pal (2013)
Student drop outs	CART	Decision tree	Yadav et al. (2011)
Student placement prediction	Naïve Bayes	Decision tree	Pal & Pal (2013)

technique hierarchical algorithms like HACM, SLINK, COBWEB, BIRCH and density based algorithms were hardly used.

Many studies should be conducted in order to understand the usage of these techniques and algorithms in the context of student performance analysis in HEIs. Such studies could throw more light to the models developed and identify if a combination of algorithms needs to be used for optimized analysis or newer algorithms need to be developed.

It can also be seen from the extant literature that researchers like Tair & El-Halees (2012) have used all the three techniques, namely association rules, classification and clustering techniques, for their research when there are some other researchers (mentioned above) have used only one of the technique. So it can be seen that there is no clarity or consensus on the techniques that were used. Furthermore, whether all of the techniques or a combination of them needs to be used, is unknown.

So the above arguments highlight the need for an investigation into data mining techniques that study the student performance.

The following sections discuss the possible factors affecting student performance, prior learning, sequence of courses and time to degree that could be investigated in this research.

3.3 Previous education history and student performance

There have been lot of studies on whether previous education history (prior learning) has any impact on the performance of student in HEIs. Researchers like Cortez & Silva (2008) have argued that students' performance is highly correlated with their performance in the previous years, and with other cultural, social and academic characteristic of the students. Researchers like Simpson (2006), Woodman (2001), Ramaswami & Bhaskaran (2010) and Bharadwaj & Pal (2012) have agreed that previous education history seems to contribute to student success. Students' prior learning attributes strongly relate to HEI outcomes while high school grades and SAT scores are known to be related to academic performance.

Researchers like Kember (1995), however, argued that characteristics of students' background are not good predictors of student performance as they are mere starting attributes and there could be more significant factors that may contribute to student performance. The predictive capability of high school preparation falls when variables demonstrating college academic performance are included in the models.

Even though there is a mixed evidence of whether previous education background (prior learning) is a significant factor that contributes to student performance or success, it implies that further investigation will help educators to enrol the correct students, and students to enrol in correct programmes. This investigation could help educators to reform the entry criteria, introduce induction courses, thereby producing high quality graduates. Secondly, the literature has paid less attention to background education as a whole; whereas most of the studies dealt with previous education grades or GPA, this study will consider previous education background as a whole including the school type, secondary or bachelors major, curriculum type, English test scores and previous education grades/GPA/percentage.

3.4 *Sequence of courses and student performance*

Much research has been done to study the significance of finishing certain gateway courses earlier in the study period in order to successfully complete the degree on time. More than 70% of students who successfully completed their bachelor's degree had completed the mathematics courses in the first two years of enrolment.

Data from a national survey (High School & Beyond [HS&B]) showed that the chance of degree completion increases by around 42% with the completion of three mathematics courses. Herzog (2006), who conducted an analysis in a large public university, found that the first year mathematics course was the second important predictor of retention after the first year GPA. He also established that first-year students who did not take mathematics courses were five times less likely to return the following year.

There are a number of studies conducted that have found that registering a mathematics course was related to the probability of transfer and degree completion. For instance, public college students who finished two mathematics courses were seen to comprise 19% of all transfers while the total credits received in college-level mathematics is believed to be a significant forecaster of degree completion.

There are few studies that suggest that completion of other courses can also act as indicators of success. Students taking a science course were related to degree success. Students who took one science course were 20% more likely to complete the degree and students who took three science courses were 27% more likely to finish. Completion of a college-level writing course increases the chances of successful completion of a bachelor's degree by 85%.

Vialardi et al. (2011) developed a recommendation system to recommend courses to be taken by students, based on the number of courses enrolled currently, courses, schedules, sections, classrooms, professors, GPA of students with similar academic yield while starting the term and ending the term, and grades obtained by students. In order to find out the courses to which the student has to be enrolled they analysed data of student who had already taken the course.

From the above literature it can be seen that enough emphasis is given to the courses to which the student has to enrol in order to achieve success or to improve performance. But, little research has been done in this area. Even though some researchers have attempted to address this they were not able to achieve attaining the sequence of courses which a student has to take in every semester or year in order to reach success in a timely manner. For example, students need to study certain compulsory subjects. These compulsory courses are classified as having importance to their major of study, college of study and university as a whole. Such courses could be studied at various points of time by the students regardless of any pre-requisite assigned to them. In such a situation, it is not possible to know which combination of courses is the best possible, that the students could study at a particular point of time. This could lead to a situation where the students are unable to know how they can perform better by taking which pattern of subjects. Additionally, such knowledge could also enable the advisers of students to advise them suitably to enhance their performance. Further, this knowledge could also enable the university to take such decisions as which courses need to be offered and when in the tenure of students so that it is possible to optimize the resources as well as make decisions that could improve the efficiency and productivity of the institution. These aspects have not been studied in depth by researchers who are involved in data mining techniques applicable to HEIs.

3.5 *Length of study and student performance*

Time-to-degree can be defined as the number of semesters (excluding summer sessions) taken towards graduation. As few as 40% of admitted students graduate with a bachelor's degree within four years of entering college. This increases to 45.7% when nine years for degree completion were tracked.

Longer time-to-degree can cost institutions space and money. If students can complete their degrees in fewer semesters, it could save money and space for the institution to bring in new students, thus making it easier to increase the percentage of graduates.

The economic downturn of the 1980s placed pressure on colleges and universities to document their effectiveness and efficiency. Comparisons can be drawn to the current post-secondary environment in the United States. By 1994, little research had been conducted on the topic of time-to-degree, but it was emerging as an important outcome.

Given the current recession and financial restraints placed on colleges, time-to-degree has re-emerged as a significant outcome for students. Numerous factors have been found to be associated with an increased time-to-degree. Utilizing Knight's version of the I-E-O framework, they can be categorized into inputs, environments, and outputs. Inputs such as gender, socio-economic status, race and preparation have all been found to have either a direct or indirect effect on time-to-degree. During college environment variables, such as credits per semester, summer enrolment, changes in major, first year GPA and continuous enrolment also have an effect on time-to-degree. Finally, end of college enrolment variables such as missed credits and total credits earned have been found to affect time-to-degree. While there is an abundance of literature on time-to-degree for all students, little is known on how time-to-degree varies within different groups, such as first-generation students who tend to have a longer time-to-degree than continuing-generation students.

As the previous studies have argued that time to degree treats it as student or institution performance, this study is unique in a way that it will include time-to-degree as directly as student performance along with GPA. There is a paucity of studies that treat GPA and time-to-degree as attributes of student performance/success. In this context some of the variables that need to be investigated include grades/percentage of marks scored by students and length of study.

Furthermore, the literature hints at a relation between previous education (prior learning), sequence of courses and time-to-degree. Moreover the literature indicates a relation between previous education (prior learning), sequence of courses and student performance (GPA). For instance Vialardi et al. (2011) have addressed the issue of finding the right course to be registered by the student based on those of the students who have already done the course. Further, the inter-relationship between time-to-degree and student performance (GPA) is addressed by a few researchers but they have used the first year GPA and have not studied the impact of the final GPA on time-to-degree. Furthermore, they have not studied in detail whether time-to-degree increases when GPA increases, that is, whether there is a positive correlation or a negative correlation between time-to-degree and student performance (GPA).

4 CATEGORIZATION OF DATA MINING ALGORITHMS

The data mining techniques used on student performance are categorized as two that are used for performance analysis and two for performance forecast. Under the first category, performance analysis, two techniques are used, namely the association technique and the clustering technique while under performance forecast classification technique and regression technique are used.

The association technique relates student performance to student demography, family background, gender, prior learning, courses and socio-economic status.

The clustering technique groups students with similar characteristics, family background, performance, demography, courses and socio-economic status.

The classification and regression techniques together predict student performance and student behaviour patterns.

5 DATA MINING

Data mining is the process used to identify functional and understandable correlations and patterns within existing huge volumes of data which cannot be done by normal query processing tools (Chung & Gray 1999). Data mining is supported by three technologies, namely large data, computers and data mining algorithms. Data mining started in 1992 when its inventor Thomas Blischok conducted a study of sales for an American drug store. He found out the correlation between sales of diapers and beer. Data mining uses basics of artificial intelligence, statistics, machine learning and advanced modelling techniques to predict future trends in business and behaviour. High-level knowledge that is not evident from raw data is revealed through data mining. The two major capabilities of data mining that generate new business chances are automated prediction of trends, behaviours and automated discovery of unknown patterns. Earlier these analyses required a lot of effort and time.

To extract meaningful knowledge from educational data, educational data mining (EDM) uses general data mining methods. In recent years EDM has developed as an autonomous area of research, which reached its peak in 2008 with the origin of the Educational Data Mining Conference, and the Journal of Educational Data Mining (Baker et al. 2008).

5.1 *Data mining methods and tasks*

Prediction and description are the main goals of data mining. Prediction forecasts unknown or future values from some variables in the database. Description is used to find patterns describing the data which can be understood by humans. Few predictive models are capable, however, of being descriptive and *vice versa*. Using data mining methods or tasks, prediction or description could be accomplished.

Classification assembles an item into one of some predefined classes (Hand 1981, Weiss & Kulikowski 1991). One of the illustrations of classification is classifying trends of finance (Apte & Hong 1996). Another example is the recognition of objects automatically from huge picture databases (Fayyad et al. 1996a).

Regression plots an item to a true-valued forecast variable. The applications of regression are various, for instance estimation of the likelihood of a patient to stay alive with the results of some diagnostic tests, and the prediction of customer requirements for a new item for consumption as a function of publicity expenditure.

Clustering is a frequently used as a descriptive task which helps to find a fixed set of clusters to explain the data (Titterington et al. 1985, Jain & Dubes 1988). Examples of clustering applications include discovering consistent sub-populations for customers in promotion databases (Cheeseman & Stutz 1996).

Different data mining algorithms have various usage and drawbacks. For instance, the linear regression algorithm is used for initial investigation of data and straight line relationships but has a drawback in that it does not allow discrete values. Association rules are used for market basket analysis but have a drawback in that they do not allow continuous values and are mainly used for recommendation systems. Each of the algorithms has its own weaknesses and strengths. It is a very challenging task to select the best algorithm for the defined problem.

Generally researchers have created data mining models by following the steps below:

- Include the column of data that might be needed after creating the underlying mining structure.
- Selection of the algorithm which is aptly suited for the analytical task.

- Select the columns and the method (prediction or input, etc.) which are to be used in the model from the structure.
- To fine-tune the processing by the algorithm parameters are set.
- By processing the structure and model, the model is populated with data.

The data mining process is iterative. The data sets used are the test and training and validation data set. The training data set is fed into the algorithm to create the primary model. In order to test if the model generated is not dependent on the training set validation data sets are used. Data other than the training sets are called holdout data. These are used to validate the model's correctness and use.

6 CONCLUSION AND FURTHER WORK

Even though there are many studies that investigate the student performance factors, various factors like sequence of courses, prior learning, time-to-degree have not been exhaustively examined using data mining. This paper demonstrated that there could be a possible linkage between the temporal sequence of courses on which a student enrols based on those of successful students from similar prior learning backgrounds and student performance (GPA) and time-to-degree. This paper also highlights that there could be a possible correlation relationship between time-to-degree and student performance (GPA). Currently the authors are working on a pilot study on the discussed research gap. In further work, the authors will find the best data mining techniques to establish the relationships, and report them in the forthcoming papers.

REFERENCES

Al-Radaideh, Q., Al-Shawakfa, M. & Al-Najjar, M. 2006. Mining student data using decision trees. Paper at *The 2006 International Arab Conference on Information Technology*. Jordan: Yarmouk University.

Anwar, M.A. & Ahmed, N. 2011. Knowledge mining in supervised and unsupervised assessment data of students' performance. *Second International Conference on Networking and Information Technology*. IPCSIT 17.

Apte, C. & Hong, S.1996. Predicting equity returns from security data. *Advances in Knowledge Discovery and Data Mining*. AAAI Press and MIT Press.

Badri, M. & Abdulla, M. 2004. Awards of excellence in institutions of higher education: An AHP approach. *International Journal of Educational Management*, 18(4): 224–242.

Baker, R.S.J.D., Barnes, T. & Beck, J.E. (eds) 2008. *Proceedings of the First International Conference on Educational Data Mining*.

Baker, R. & Yacef, K. 2009. The state of educational data mining in 2009: A review and future visions. *J. Educ. Data Mining* 1(1): 3–17.

Becket, N. & Brookes, M. 2006. Evaluating quality management in university departments. *Quality Assurance in Education* 14(2): 123–142.

Becket, N. & Brookes, M. 2008. Quality management practice in higher education: What quality are we actually enhancing? *Journal of Hospitality Leisure Sport and Tourism Education* 7(1): 40–54.

Bharadwaj, B.K. & Pal, S. 2011. Mining educational data to analyze students' performance. *International Journal of Advanced Computer Science and Applications* 2(6): 63–69.

Bharadwaj, B.K. & Pal, S. 2012. Data mining: A prediction for performance improvement using classification. *International Journal of Computer Science and Information Security* 9(4): 136–140.

Borahan, N.G. & Ziarati, R. 2002. Developing quality criteria for application in the higher education sector in Turkey. *Total Quality Management* 13(7): 913–926.

Castellano, E., Martínez, L. 2008. 'ORIEB, A CRS for academic orientation using qualitative assessments', *Proceedings of the IADIS International Conference E-Learning*.

Cheeseman, P. & Stutz, J. 1996. Bayesian classification (AUTOCLASS): Theory and results. *Advances in Knowledge Discovery and Data Mining*.

Cheng, Y. & Tam, W. 1997. Multi-models of quality in education. *Quality Assurance in Education* 5(1): 22–31.

Chung, H.M. & Gray, P. 1999. Special section: Data mining. *Journal of Management Information Systems* 16(1): 11–17.

Cortez, P., Silva, A. 2008. Using data mining to predict secondary school student performance. *Proceedings of Fifth Future Business Technology Conference*. Oporto, Portugal.

Dekker, G., Pechenizkiy, M. & Vleeshouwers, J. 2009. Predicting students drop out: A case study. *Proceedings of the Second International Conference on Educational Data Mining (EDM'09)*. Cordoba, Spain.

Dew, J.R. 2009. Quality issues in higher education. *Journal for Quality and Participation* 32(1).

Eriksen, S.D. 1995. TQM and the transformation from an elite to a mass system of higher education in the UK. *Quality Assurance in Education* 3(1): 14–29.

Fayyad, U.M., Djorgovski, S.G. & Weir, N. 1996a. From digitized images to on-line catalogs: Data mining a sky survey. *AI Magazine* 17(2): 51–66.

Fayyad, U., Piatetsky-Shapiro, G. & Smyth, R. 1996b. The KDD process for extracting useful knowledge from volumes of data. *Communications of the ACM* 39(11): 27–34.

Hand, D.J. 1981. *Discrimination and Classification*. Chichester, U.K.: Wiley.

Hénard, F. 2008. Learning our lesson: Review of quality teaching in higher education. *Institutional Management in Higher Education*.

Herrera, O.L. 2006. *Investigation of the Role of Pre- and Post-admission Variables in Undergraduate Institutional Persistence using a Markov Student-flow Model*. Ph.D. Dissertation, North Carolina State University, USA.

Herzog, S. 2006. Estimating student retention and degree-completion time: Decision trees and neural networks vis-à-vis regression. In J. Luan & C. Zhao (eds), *New Direction for Institutional Research* no. 131. San Francisco: Jossey-Bass.

Huebner, R.A. 2013. A survey of educational data-mining research. *Research in Higher Education Journal*.

Jain, A.K. & Dubes, R.C. 1988. *Algorithms for Clustering Data*. Englewood Cliffs, N.J.: Prentice-Hall.

Kember, D. 1995. *Open Learning Courses for Adults: A Model of Student Progress*. Englewood Cliffs, NJ: Education Technology.

Kotsiantis, S., Pierrakeas, C. & Pintelas, P. 2004. Predicting students' performance in distance learning using machine learning techniques. *Applied Artificial Intelligence* 18: 411–426.

Kowalkiewicz, A. 2007. The impact of quality culture on quality of teaching: A case of business higher education in Poland. *First European Forum for Quality Assurance*.

Kumar, A.S. & Vijayalakshmi, M.N. 2011. Efficiency of decision trees in predicting students' academic performance. *CCSEA, CS & IT* 02: 335–343.

Luan, J. 2001. Data mining and knowledge management: A system analysis for establishing a tiered knowledge management model (TKMM). *Proceedings of AIR Forum, Toronto, Canada*.

Luan, J. 2002a. Data mining and knowledge management in higher education: Potential applications. *Proceedings of AIR Forum, Toronto, Canada*.

Luan, J. 2002b. *Data Mining Application in Higher Education*. SPSS Executive Report.

Lykorentzou, I., Giannoukos, I., Nikopoulos, V., Mpardis, G. & Loumos, V. 2009. Dropout prediction in e-learning courses through the combination of machine learning techniques. *Computers & Education* 53: 950–965.

McKay, J. & Kember, D. 1999. Quality assurance systems and educational development. Part 1: The limitations of quality control. *Quality Assurance in Education* 7(1): 25–29.

Martens, E. & Prosser, M. 1998. What constitutes high quality teaching and learning and how to assure it. *Quality Assurance in Education* 6(1): 28–36.

Mizikaci, F. 2006. A systems approach to programme evaluation model for quality in higher education. *Quality Assurance in Education* 14(1): 37–53.

OECD 1994. *Quality in Teaching*. Paris.

Oldfield, B. & Baron, S. 1998. Is servicescape important to student perceptions of service quality? *Research Paper*, Manchester Metropolitan University.

Osmanbegović, E. & Suljić, M. 2012. Data mining approach for predicting student performance. *Economic Review – Journal of Economics and Business* X(1) [10(1)].

Owlia, M. & Aspinwall, E. 1996. A framework for the dimensions of quality in higher education. *Quality Assurance in Education* 4(2): 12–20.

Pal, K. & Pal, S. 2013. Classification model of prediction for placement of students. *IJMECS* 5(11): 49–56.

Quadril, M.N. & Kalyanka, N.V. 2007. Drop out feature of student data for academic performance using decision trees. *Global Journal of Computer Science and Technology* 10(2) (Ver 1.0).

Ramaswami, M. & Bhaskaran, R. 2010. A CHAID-based performance prediction model in educational data mining. *Int. J. Comput. Sci. Issues* 7(1): 10–18.

Ramsden, P. 1992. *Learning to Teach in Higher Education.* New York: Routledge.

Romero, C. & Ventura, S. 2007. Educational data mining: A survey from 1995 to 2005. *Expert Syst. Appl.* 1(33): 135–146.

Simpson, O. 2006. Predicting student success in open and distance learning. *Open Learning* 21(2): 125–138.

Siraj, F. & Abdoulha, M.A. 2009. Uncovering hidden information within universities' student enrolment data using data mining. *MASAUM Journal of Computing* 1(2): 337–342.

Titterington, D.M., Smith, A.F.M. & Makov, U.E. 1985. *Statistical Analysis of Finite-Mixture Distributions.* Chichester, U.K.: Wiley.

Tair, M.M.T. & El-Halees, A.M. 2012. Mining educational data to improve students' performance: A case study. *International Journal of Information and Communication Technology Research* 2(2).

Thai Nge N., Janecek, P. & Haddawy P. 2007. A comparative analysis of techniques for predicting academic performance. 37th *ASEE/IEEE Frontiers in Education Conference.*

Vandamme, J.-P., Meskens, N. & Superby, J.-F. 2007. Predicting academic performance by data mining methods. *Education Economics* 15(4): 405–419.

Vialardi, C., Chue, J., Barrientos, A., Victoria, D., Estrella, J., Ortigosa, A. & Peche, J. 2011. A data mining approach to guide students through the enrolment process based on academic performance. *User Modeling and User-Adapted Interaction* 21(1–2).

WEA (World Education Forum) 2000. The Dakar Framework for Action. Paris: UNESCO.

Waiyamai, K. 2003. Improving Quality of Graduate Students by Data Mining. Dept Computer Engineering, Faculty of Engineering, Kasetsart University, Bangkok, Thailand.

Weiss, S.I. & Kulikowski, C. 1991. *Computer Systems That Learn: Classification and Prediction Methods from Statistics, Neural Networks, Machine Learning and Expert Systems.* San Francisco, CA: Morgan Kaufmann.

Woodman, R. 2001. Investigation of factors that influence student retention and success rate on Open University courses in the East Anglia region. M.Sc. Dissertation, Sheffield Hallam University, U.K.

Yadav, S.K., Bharadwaj, B.K. & Pal, S. 2011. Data mining applications: A comparative study for predicting students' performance. *International Journal of Innovative Technology and Creative Engineering (IJITCE)* 1(12): 13–19.

Yorke, M. 1999. Assuring quality and standards in globalised higher education. *Quality Assurance in Education* 7(1): 14–24.

Yu, C.H., DiGangi, S., Jannasch-Pennell, A., Lo, W. & Kaprolet, C. 2007. A data-mining approach to differentiate predictors of retention. *Proceedings of the Educause Southwest Conference, Austin, Texas, USA.*

Organizational effectiveness in secondary schools: An empirical study

Tahani H.A. Maki
Ahlia University, Kingdom of Bahrain

Satwinder Singh & Tillal Eldabi
Brunel University London, UK

Wajeeh Elali
Ahlia University, Kingdom of Bahrain

ABSTRACT: This paper outlines a research plan to measure the performance of secondary schools (SS). Following a literature review, it is argued that the performance of SS measured with the help of a composite index of students' academic achievements, students' personal development, quality and effectiveness of teaching, learning and curriculum implementation, support and guidance for students, and the quality and effectiveness of leadership, management and governance, can be measured by taking account of internal and external factors impacting on performance. Internal factors include growth plans, implicit and explicit powers held by Head of School (HoS), organization structure, leadership style, and HR policies. External factors include demand for organizations' services, competitive conditions, political, economic, social, legal and technical environment. We present a conceptual model to be empirically tested along these lines with theoretical underpinnings from the leader-member exchange theory (LMX) and the principal-agent model (PAM).

1 INTRODUCTION

This conceptual paper outlines the approach to be adopted for an empirical study to examine the organizational effectiveness (OE) of secondary schools (SS). It is not yet known why these differences exist and as yet no study has been undertaken to systematically analyse the reasons behind the differential performance of SS on the parameters mentioned. This study is aimed at filling this gap.

The study assumes importance on three grounds. First, by identifying and isolating the factors that contribute to differential performances, the SS administrators would be able to take corrective actions to improve their schools' performances. Second, the study will help government adopt policy measures to improve government-assisted SS resulting in better returns for government's investment in public schools. Third, the study, will pave the way for future academic research.

The paper is organized as follows. Section 2 describes issues concerning OE. A literature review occupies section 3. Section 4 outlines the research question, aims and objectives. A framework for measuring OE is presented in section 5. Theoretical underpinnings of the study are explained in section 6. Data and measurement techniques are described in section 7.

2 THEORETICAL ISSUES CONCERNING ORGANIZATIONAL EFFECTIVENESS

2.1 *Organizational effectiveness*

Measuring organizational effectiveness (or performance, as it is sometimes referred to) is a difficult task. The difficulty arises from the fact that there is, as yet, no consensus among researchers on how precisely to measure OE (Scott 1977). Conceptually, for a for-profit organization, OE can be defined as the comparison of the value produced by a company with the value owners expected to receive from the company (Alchian & Demsetz 1972). Organizational effectiveness can be also be defined in terms of HRM-related outcomes, such as turnover, absenteeism, job satisfaction, commitment, or various organizational outcomes, such as productivity, quality, service, efficiencies, customer satisfaction (Dyer & Reeves 1995). Furthermore, it can also be defined in terms of financial indicators, i.e. profits, sales, return on assets, equity, or investment, or capital market outcomes – market share, Tobin's q, stock price, and growth. In the business world, the focal point of attention on this construct has been almost completely directed towards financial measures of performance (Rowe et al. 1995). For not-for-profit organizations, the task of defining OE assumes further challenges as the nature, characteristics, and functioning of such organizations differ widely. The present study focusses on the education sector within services.

2.2 *Measuring effectiveness in secondary schools*

In the case of SS, attempts have been made by authors to define performance. Secondary school performance may be regarded as the integration of students' characteristics, structural characteristics (public *versus* private schools), school resources and school processes.

Authors have added further perspectives to this concept of SS performance. The survey by Rockoff et al. (2011) also encompasses non-traditional measures like personality traits, feelings of self-efficacy and cognitive ability. Lydiah & Nasongo (2009) used the improvement measures applied by school principals' and teachers' indulgence in school management through team work to examine their influence on the performance of secondary school in terms of students' academic performance (Lydiah & Nasongo 2009, Musera et al. 2012). Researchers have also defined school performance in terms of teachers' performance (Mohammad et al. 2011), students' academic performance (Irfan & Shabana 2012, Farooq et al. 2011) and school principal leadership styles (Wan & Jamal 2012, Beare et al. 1989, Sammons et al. 2011, Jacobson 2011) and external elements such as parents' involvement (this includes behaviour practices such as parental aspirations, expectations, attitudes and beliefs related to students' education, direct and indirect longitudinal effects of parental involvement in students' achievement (Hoover-Dempsey et al. 2001, Pamela & Davis-Keen 2005) and educational policies (Glatter 2002).

For the purpose of this study, OE of SS in Bahrain is measured using a composite index of academic achievements, students' personal development, the quality and effectiveness of teaching, learning and curriculum implementation, support and guidance for students, and the quality and effectiveness of leadership, management and governance.

3 LITERATURE REVIEW

3.1 *Relevant literature*

Researchers have investigated secondary schools' performance in relation to such variables as leader member exchange relations (Graen & Uhl-Bien 1995, Glynn & DeJordy 2010), job characteristics (Hogg et al. 2005, Wells & Feun 2007, Lee 2008, Atwater & Carmeli 2009, Clemens et al. 2009). In the literature, teachers have been identified as the significant factor affecting OE. Research suggests that teacher performance (impacting on OE) is linked to different practices, including teamwork (Wichenje et al. 2012, Riketta & van Dick 2005), working conditions that include school administration support, salaries and overcoming of students' discipline problems (Ingersoll 2001). Researchers have also linked secondary school performance with students' performance that is affected by factors related to school facilities such as classroom size, technology used in the classroom and school culture (Kythreotis et al. 2010, Sammons et al. 2011, Irfan & Shabana 2012). The third set of factors relate to admin factors including leadership style (Mudalia 2012), leadership skills (Orora 1997) and involvement in decision making (Koontz & Weinhrich 1998), teamwork and delegation (Koontz and Weinhrich 1998). Some researchers have also attributed performance to external factors including a family's social, educational and economic background (Caldas & Bankston 1997, Jeynes 2002, Parelius & Parelius 1987, Mitchell & Collom 2001, Ma & Klinger 2000).

Researchers have also investigated variables related to gender differences (Kettle 1997, Fennell 1999). Grogran (1999) found that women view the position of school principal as an educational leader, while men view it as a managerial position (Manjulika et al. 2008). Some researchers have, however, linked school performance with leadership support of teachers. For instance, Fiore (2004) found that the absence of affective leadership leads teachers to leave their job.

Another piece of research has studied the impact of secondary school principals' job satisfaction and motivation on the performance of schools. The study found that secondary school principal job satisfaction is affected by external factors such as continual curriculum changes, government policies, school structure and school budget (Chaplain 2001, Maforah & Schulze 2012). Previous literature on secondary schools indicated that teachers' teamwork is one way to enhance the quality of secondary school performance. Mondy & Robert (1981), Dubrin (1981) and Robbins (1982) have emphasized that teachers' teams play a vital role in providing counselling, training, data feedback to teachers, improve schools performance, achieving schools' goals in the specified time and provide clear communication. Madiha (2012) found that collegiality could affect secondary school teachers' and students' performance, which also could impact the organization's commitment. A major challenge for teachers' teams in secondary schools is, however, the performance appraisal. Chance (1989) has suggested that the school should consider the team as one unit when providing them with the appraisal in order to encourage co-operation among teachers' team members.

A number of studies have also examined the constraints on secondary schools' effectiveness. Renlhan & Renlhan (1984), Murphy (1987) and Goodlad (1990) found the following factors to be constraints on school-effectiveness: school size and complexity, school purpose, nature of the school, procedures and routines, elements of school workforce, school strategies, technology selection, weak leadership, life cycle stage, educational support and tradition. Leadership effectiveness is affected by the following: staff mobility, time management in solving students' problems and management tasks and shortage of facilities. Cuban (1984), Ralph & Fennessey (1983) concluded that there is no consensus related to the definition of effective schools. Renlhan & Renlhan (1984) and Duren (1992), however, identified the following elements of school effectiveness: stability of staff, recognition of academic success, community and parental support and articulation of the curriculum. Other researchers have suggested several culture variables that are required for effectiveness of an organization such as provision of a safe environment, collaboration in planning, community support, clarification of goals, maintenance of discipline, student involvement, creation of high expectations for achievement and building of a specific organizational culture (Fullan 1982, Johnson 1988). Smith & Holdaway (1995) in their study conclude that school effectiveness is impeded by financial constraints, leadership constraints, time constraints and staffing constraints.

Burnett et al. (1999), Ames & Ames (1989), Elliot & Dweck (1988) and Maeher & Anderman (1993) have linked secondary school effectiveness with the concept of performance-focussed goals, that is, the focus on the ability of students, teachers and principal; this also includes getting better grades, getting rewards, approval and incentives from others (Maeher & Anderman 1993, Midgley (1993), Midgley et al. 1995). It was found that students' performance is affected by teachers' organizational commitment (Marks & Louis 1997).

3.2 Mediating factors

The impact of mediating factors on performance has also been pointed out in the literature (Liden & Maslyn 1998, Kraimer & Wayne 2004, Hogg et al. 2005, Kennedy et al. 2012). As a result of this, some researchers have started studying these mediating variables to understand which of them could impact secondary school performance to the greatest extent (Eisenberger et al. 1986, Meyer et al. 1989, Eisenberger et al. 1990, Kouzes & Posner 1995, Pounder et al. 1995, Greguras & Ford 2006, Somech & Wenderow 2006). Mediating variables studied include commitment, empowerment, satisfaction, trust, motivation, absenteeism, turnover and burnout (Begley & Czajka 1993, Tharenou 1993, Dee et al. 2006, Balyer 2012).

Research taking into account the mediating factors is based on the idea that secondary schools performance is contingent upon the teacher, principal, student practices and other aspects of schools such as the learning environment, human resources management (HRM), and the working environment (Kennedy et al. 2012). Hallinger & Hech (1998) found that the relationship between school leadership and student performance is mediated by several variables that include a school's purposes and goals, school structure and school culture. As a result, a group of school practices and different measures have been identified by researchers. It has also been suggested that variables that could have mediated, for example, the relationship between practices of school leadership and school performance might have a different impact on the performance of students, teachers and the school in general (Naseer 2011).

3.3 *Gaps in the literature*

A review of the literature reveals that, although research in schools' effectiveness has been undertaken, there is, as yet, no consensus amongst researchers on factors that affect secondary school performance (Hallinger & Heck 1996a, b, Heck et al. 1990, Bryk & Raudenbush 1992). It has been pointed out that results obtained thus far cannot be generalized because different factors might affect the educational performance of individual secondary schools (Eghbal et al. 2011). In other words, although there might be a common set of factors impacting on OE, certain factors might be school-specific. One limitation of research conducted thus far has been that performance measures are mostly focussed on academic performance criteria, with few studies taking into consideration other issues related to secondary school performance such as social and psychological issues (Wan & Jamal 2012). There is also disagreement among researchers about the method of data collection (Jagero 2011, Eghbal et al. 2011).

A further review of the literature showed that studies conducted thus far on the performance of schools have been conducted with the help of a selected number of variables and these studies have not accounted for aspects of performance such as teamwork and its associated variables such as the role of the school principal, team members or followers, motivation, trust, job characteristics, work considered meaningful or not, and teachers' motivation and performance. The suggested framework in the following section attempts to develop an approach to measuring the performance of secondary schools in the light of the different proposed variables of teachers' teamwork. The studies thus far have also left out external factors some of which are shown in Figure 1 (Singh et al. 2012).

4 RESEARCH QUESTION, AIMS AND OBJECTIVES

4.1 *Research question*

A research question arising from this study of the literature is, "Can trust act as a mediator between the leader, followers in the team and the overall performance of teachers' team work?"

4.2 *Aim of the research*

The associated aim of the research is to study the relationship between the leader and members of a team of teachers in order to understand how trust among them affects this relationship.

4.3 *Objectives*

The associated objectives are

- To develop hypotheses related to the relationship between the leader and followers of teachers' teams mediated by trust.

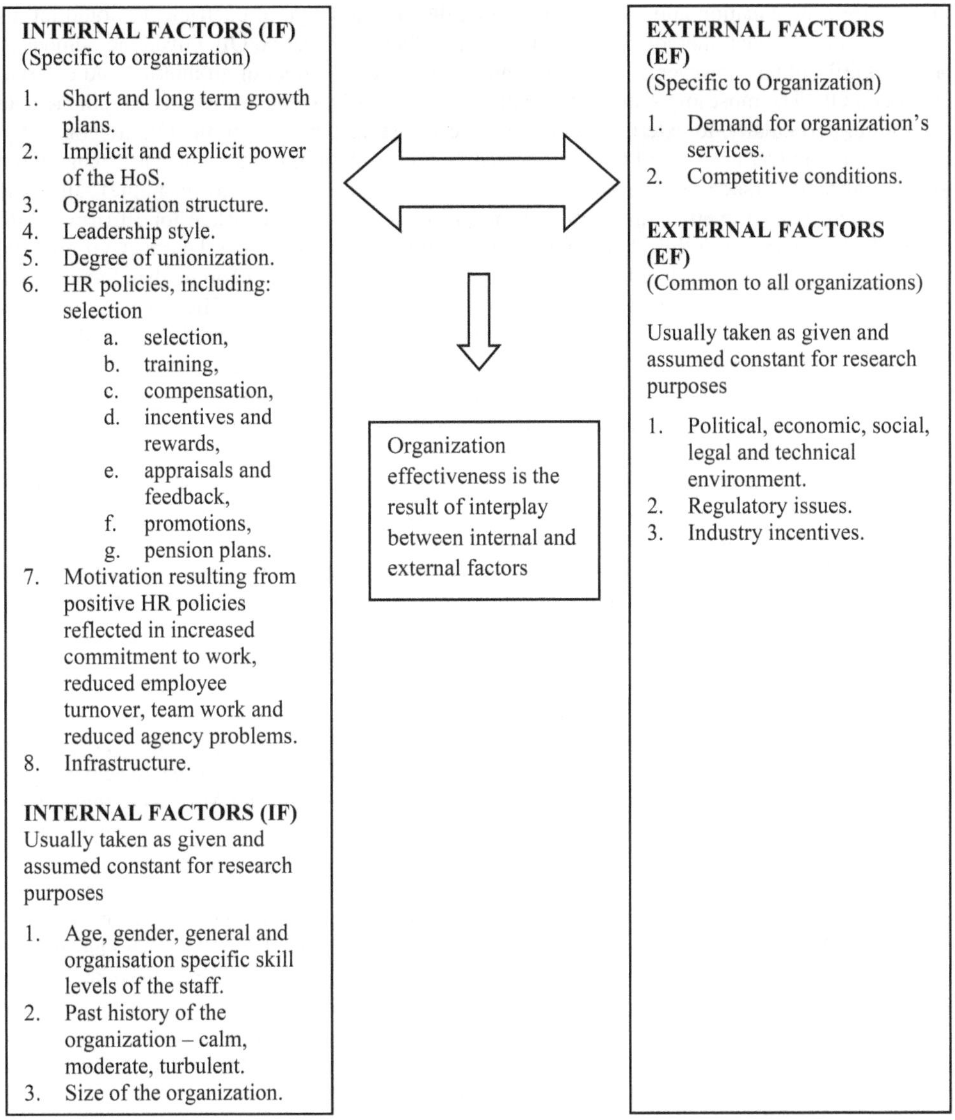

INTERNAL FACTORS (IF)
(Specific to organization)

1. Short and long term growth plans.
2. Implicit and explicit power of the HoS.
3. Organization structure.
4. Leadership style.
5. Degree of unionization.
6. HR policies, including: selection
    a. selection,
    b. training,
    c. compensation,
    d. incentives and rewards,
    e. appraisals and feedback,
    f. promotions,
    g. pension plans.
7. Motivation resulting from positive HR policies reflected in increased commitment to work, reduced employee turnover, team work and reduced agency problems.
8. Infrastructure.

INTERNAL FACTORS (IF)
Usually taken as given and assumed constant for research purposes

1. Age, gender, general and organisation specific skill levels of the staff.
2. Past history of the organization – calm, moderate, turbulent.
3. Size of the organization.

Organization effectiveness is the result of interplay between internal and external factors

EXTERNAL FACTORS (EF)
(Specific to Organization)

1. Demand for organization's services.
2. Competitive conditions.

EXTERNAL FACTORS (EF)
(Common to all organizations)

Usually taken as given and assumed constant for research purposes

1. Political, economic, social, legal and technical environment.
2. Regulatory issues.
3. Industry incentives.

Figure 1. A schematic interplay of variables impacting OE (Singh et al. 2012).

- To examine the impact of the relationship between leader, followers and trust on the performance of teachers' teams.
- To study the literature of leader member exchange to gain understanding of the different mechanisms of team work.

5 A FRAMEWORK FOR MEASURING ORGANIZATIONAL EFFECTIVENESS

Figure 1 schematically displays the interplay of factors that would have an impact on OE in the context of SS. In an ideal situation, in order to accurately measure the impact of OE, the individual effect of all the internal and external factors, as listed in Figure 1, should be taken into account.

In a multivariate equation approach, for example, the size and sign of coefficients would tell us the importance, or lack thereof, associated with each factor's effect on OE. Given the number of practical difficulties, however, it is not possible to measure the effect of all internal and external factors on OE. The most important of these difficulties, clearly, is the availability of reliable data pertaining, particularly, to external factors. As a result, studies relating to the OE are conducted under the implicit assumption of the *ceteris paribus* clause. What this clause means is that, owing to external factors (such as the macro-economic policies of a country, for example), all firms are impacted in a symmetric way, with their overall impact taken as constant for all firms. With the implicit declaration of this clause, researchers can focus on measuring the effects of only the internal HR practices and related factors on OE. In this regard, one could argue that, by doing so, researchers implicitly subscribe to the notion that measuring the effect of HR practices on OE is an inexact science, but, because all firms are influenced in a similar manner by external factors, if the effect of *all* the internal factors on OE can be captured, the results can provide a fairly good measure of the impacts of HR practices on OE. In the course of literature review and hypothesis-building stage for this work it was discovered that most studies conducted on OE in the context of SS have not been able to take into account all internal factors (listed in Fig. 1). Secondly, in order for the results to be accurate, all firms in a particular sector should be accounted for, which often is not the case. Thirdly, the literature review also revealed that only a limited number of controls (firm size, turnover, etc.) have been included in the studies. A final associated point concerns the measurement of outcome measured for OE. Most studies take into account just one measure (e.g., the pass rate of students). In this study we intend to address all these lacunae in the empirical work.

6 THEORETICAL UNDERPINNINGS OF THE STUDY

Although the aim of this study is to take a system's view, the theoretical underpinnings are embedded in the Leader-member exchange (LMX) theory (Gerstner & Day 1997) and the Principal-Agent model (PAM) (Jensen & Mackling 1976, Mitnick 1974, Ross 1973). The LMX model focusses on the two-way relationship between supervisors and their subordinates. Theory assumes that leaders develop an exchange with each of their subordinates, and that the quality LMX relationships influence subordinates' responsibility, decision influence, access to resources and performance (Deluga 1998, Linden et al. 1997). Somewhat allied to this notion is the PAM. In its simplest form in the PAM, the principal delegates assignments and decisions to an agent who has private information about his abilities, potential and actual effort that can be exerted, and the outcome of a stochastic process. The two parties (the P and the A) have conflicting goals and possibly different risk attitudes. In this situation, the economic problem is to design a contract which maximizes the principal's expected utility given that the agent will maximize his own expected utility and would not accept a contract which offers less than his best alternative.

The core issue of asymmetric information and inability of principal to monitor the efforts of his agent lends itself to uncertainty and risk within the PA relationship and often ends up in agency costs. The solution to this problem is closely related to the moral hazard issue and the designing of appropriate incentive contracts. The problem (as addressed by game theorists) is often compounded by the nature of the contractual relationship, that is, whether it is short term or long term involving, respectively, one-off, a few, or repeated interaction between contracting parties. The availability of information and the nature of contract, for example, can decide how the agents would be compensated by way of salaries, perks, bonuses (discretionary or otherwise), promotions, and such other payments (Doeringer & Piore 1971). The issue of compensation is further compounded in the case of teams in which it is not possible to ascertain the amount of effort exerted by individual team members (Alchian & Demsetz 1972, Holmstrom 1982, McLaughlin 1994).

The presence of all these issues has led researchers to suggest optimal contract designs (Holmstrom 1979). Holmstrom's model suggests that contract design and payoffs to agents should be based on the level of four principles – informativeness principle (amount of information available to assess the effort), incentive-intensity principle (degree and responsiveness of agent to efforts and risk), monitoring assessing principle (degree of monitoring required), and equal compensation principle (activities when equally valued by the employer should be equally valued by the employees). It should also be noted that the adoption of tournament approach further complicates the incentive and compensation issues in organizations (Doeringer & Piore 1971, Rosen 1982, Lazear & Rosen 1981, Green & Stokey 1983, Carmichael 1983).

The study intends to explore various aspects of LMX and PAM in its research. Related issues such as those of trust, commitment, and motivation, will also be explored. An important aspect which is typical of SS employment is the one known in the literature as "deferred compensation" (implicitly resulting from tournaments) under which the compensation is structured in such a way that staff get larger payments when they are well advanced in their employment contracts (Salop & Salop 1976). This can be construed as a form of forced savings or as a signal for personal development (Loewenstein & Sicherman 1991, Frank & Hutchens 1993, Akerlof & Katz 1989). It is not yet known what impact this might have on organizational effectiveness.

7 DATA AND MEASUREMENT TECHNIQUES

In order to achieve the objectives of the research it is important to specify the data collection method that will be used to examine trust, the leader and followers in teachers' team work. Various researchers have addressed this topic by different methodologies. In order to achieve the aim of this research in collecting sizable data from both the leader and followers of teachers' teams, it may, however, be important to use quantitative research methodology. This is supported by several researchers such as Celik et al. (2011). A selected set of open-ended interviews will be conducted on managers and staff to clarify and add value to empirical results obtained. The statistical methodology will include the use of descriptive and multivariate techniques. The latter would include the use of zero and first order correlations, multiple regression analysis (MRA) including ordered probit and multinomial regression approaches to data analysis. At this stage it is not decided if recourse would be taken to the use structural equation modelling. If need be this technique will be used to supplement results from MRA.

8 SUMMARY

Measuring OE is a complex task, particularly for the not-for-profit services sector such as the secondary school sector where profitability or such other market measures have little meaning in assessing the performance of organizations, in this case education providing institutions. The issue though is of huge importance as large scale public funding is invested in SS and the State would wish better return for its investment. For parents the issue assumes importance as SS are followed by an entry into the university sector which is highly competitive. Schools provide foundation level skills to the young human capital, the quality of which can determine the overall skill index of a nation *vis-à-vis* other nations.

A survey of the literature showed that research on the OE of SS has been studied by scholars in different national contexts, usually taking into account such variables as school structure, students' performance (that is, pass rate and dropout rate), teachers, leadership and school management (that is, leadership style and teamwork) and school environment. In this paper we have proposed that a study on OE of SS be conducted with the help of primary and secondary data taking into account the internal and external factors impacting on performance. Suggested internal factors include growth plans, implicit and explicit powers held by Heads of School, organization structure, leadership style, and HR policies. External factors include demand for organizations' services, competitive conditions, and political, economic, social, legal and technical environments.

We have presented a conceptual model to be tested with the help of primary and secondary data on these lines with the theoretical underpinnings from the LMX and PAM including the contract design and payoffs to agents on the level of four principles – informativeness principle (amount of information available to assess the effort), incentive-intensity principle (degree and responsiveness of agent to efforts and risk), monitoring assessing principle (degree of monitoring required), and equal compensation principle (activities when equally valued by the employer should be equally valued by the employees). In the empirical analysis, the statistical methodology will include the use of descriptive and multivariate techniques. The latter would include the use of zero and first order correlations, multiple regression analysis (MRA) including ordered probit and multinomial regression approaches to data analysis.

REFERENCES

Akerlof, G.A. & Katz, L.F. 1989. Workers' trust funds and the logic of wages profiles. *Quarterly Journal of Economics* 104(3): 525–536.

Alchian A.A. & Demsetz, H. 1972. Production, information costs and economic organization. *American Economic Review* 62: 777–795.

Ames, C. & Ames. 1989. *Research on Motivation in Education: Goals and Cognitions.* New York, NY: Academic Press.

Atwater, L. & Cameli, A. 2009. Leader-member exchange, feelings of energy and involvement in creative work. *The Leadership Quarterly* 20(3): 264–275.

Balyer, A. 2012. Transformational leadership behaviors of school principals: A qualitative research based on teachers' perceptions. *International Online Journal of Educational Sciences* 4(3): 581–591.

Beare, H.C., Brian, J. & Millikan, R.H. 1989. *Creating an Excellent School: Some New Management Techniques.* New York, NY: Routledge Education.

Begley, T.M. & Czajka. J.M. 1993. Panel analysis of the moderating effects of commitment on job satisfaction, intern to quit and health following organizational change. *Journal of Applied Psychology* 78: 552–556.

Bryk, A.S. & Raudenbush, S.W. 1992. *Hierarchical Linear Models: Applications and Data Analysis Methods.* Newbury Park, CA: Sage.

Burnett, K., John, M. & Robert, K. 1999. Transformational leadership in schools: Panacea, placebo or problem? *Journal of Educational Administration* 39(1): 24–46.

Caldas, S.J. & Bankston, C.L. 1997. The effect of school population socioeconomic status on individual student academic achievement. *Journal of Educational Research* 90: 269–277.

Carmichael, L. 1983. Firm specific human capital and promotion leaders. *Bell Journal of Economics* 14: 251–258.

Celik, M, Turunc, O. & Begenirbas, M. 2011. The role of organizational trust, burnout and international deviance for achieving organizational performance. *International Journal of Business and Management* 3(2): 179–189.

Chance, P. 1989. Great experiments in team chemistry. *Across the Board* 26(5): 18–28.

Chaplain, R.P. 2001. Stress and job satisfaction among primary head teachers: A question of balance. *Educational Management and Administration* 29: 197–215.

Clemens, E.V., Milsom, A. & Cashwell, C.S. 2009. Using leader-member exchange theory to examine principal-school counsellor relationships, school councilors' roles, job satisfaction, and turnover intentions. *Professional School Counselling* 13: 75–86.

Cuban, L. 1984. Transforming the frog into a prince: Effective schools research, policy and practice at the district level. *Harvard Educational Review* 54(2): 129–151.

Dee, J.R., Henkin, A.B. & Singleton, C.A. 2006. Organizational commitment of teachers in urban schools: Examining the effects of team structures. *Urban Education* 41: 603–627.

Deluga, R.J. 1998. Leader-member exchange quality and effectiveness ratings: The role of subordinate-supervisor conscientiousness similarity. *Group and Organisation Management* 23: 189–216.

Doeringer, P.B. & Piore, M.J. 1971. Internal Labour Markets and Manpower Analysis. Lexington, MA: Heath.

Dubrin, A.J. 1981. *Personnel and Human Resources Management.* New York, NY: van Nostrand.

Duren, O.R. 1992. A reflection of my school: One student looks at school effectiveness. *NASSP Bulletin* 76(542): 93–96.

Dyer, L. & Reeves, T. 1995. Human resource strategies and firm performance: What do we know and where do we need to go? *International Journal of Human Resource Management* 6: 656–670.

Eghbal, Z., Hossein, Z. & Alireza, F. 2011. Examining the internal school factors contributing to predict the academic performance of the high school students in Hormozgan. *Journal of Educational and Management Studies* 1(1): 1.

Eisenberger, R., Fasolo, P.M. & Davis-LaMastro, V. 1990. Effects of perceived organizational support on employee diligence, innovation, and commitment. *Journal of Applied Psychology*, 53: 51–59.

Eisenberger, R., Huntington, R., Hutchison, S. & Sowa, D. 1986. Perceived organizational support. *Journal of Applied Psychology* 71: 500–507.

Elliot, E. & Dweck, C. 1988. Goals: An approach to motivation and achievement. *Journal of Personality and Social Psychology* 54: 5–12.

Farooq, M.S., Chaudhr, A.H., Shafiq, M. & Berhanu, G. 2011. Factors affecting students' quality of academic performance: A case of secondary school level. *Journal of Quality and Technology Management* VII(II): 1–14.

Fennell, H.A. 1999. Power in principalship: Four women's experiences. *Journal of Educational Administration* 37(1): 23–49.

Fiore, D.J. 2004. *Introduction to Educational Administration: Standard Theories and Practice*. Larchmont, NY: Eye on Education.

Frank, H.R. & Hutchens, R.M. 1993. Wages, seniority and the demand for rising consumption profiles. *Journal of Economic Behaviour and Organisation* 21: 251–276.

Fullan, M. 1982. The meaning of educational change. *Ontario Institute of studies in Education*, Toronto.

Glatter, R. 2002. Governance, autonomy and accountability. In T. Bush & L. Bell (eds), *The Principles and Practice of Educational Management*. London: Paul Chapman.

Glynn, M.A. & DeJordy 2010. Leadership through an organisation behaviour lens. In N. Nohria & R. Khurna (eds), *Handbook of Leadership Theory and Practice*. Cambridge, Mass.: Harvard Business Press.

Goodlad, J.M. 1990. Better teachers for our nation's schools. *Phi Delta Kappa*, November: 185–194.

Graen, G.B. & Uhl-Bien, M. 1995. Relationship-based approach to leadership: Development of leader-member exchange (LMX) theory of leadership over 25 years: Applying a multi-level multi-domain perspective. *The Leadership Quarterly* 6(2): 219–247.

Green, J. & Stokey, N. 1983. A comparison of tournaments and contracts. *Journal of Political Economy* 91: 394–364.

Greguras, G.J. & Ford, J.M. 2006. An examination of the multidimensionality of supervisor and subordinate perceptions of leader – member exchange. *Journal of Occupational and Organisational Psychology* 79(3): 433–465.

Grogan, M. 1999. Equity/quality issues of gender, race, and class. *Educational Administration Quarterly* 35(4): 518–536.

Hallinger, P. & Heck, R.H. 1996a. The principal's role in school effectiveness: An assessment of methodological progress, 1980–1995. In K. Leithwood & P. Hallinger (eds), *International Handbook of Educational Leadership and Administration*: 723–783. Dordrecht: Kluwer Academic Publishers.

Hallinger, P. & Heck, R.H. 1996b. Reassessing the principal's role in school effectiveness: A review of empirical research, 1980–1995. *Educational Administration Quarterly* 32(1): 5–44.

Hallinger, P. & Heck, R.H. 1998. Exploring the principal's contribution to school effectiveness, 1980–1995. *School Effectiveness and School Improvement* 9: 157–191.

Heck, R.H., Larsen, T.J. & Marcoulides, G.A. 1990. Instructional leadership and school achievement: Validation of a causal model. *Educational Administration Quarterly* 26: 94–125.

Hogg, M.A., Martin, R., Epitropaki, O., Mankact, A., Svensson, A. & Weeden, K. Effective leadership in salient groups: Revisiting leader-member exchange theory from the perspective of the social identity theory of leadership. *Personality and Social Psychology Bulletin* 31(7): 991–1004.

Holmstrom, B. 1979. Moral hazard and observability, *Bell Journal of Economics* 10(1): 74–91.

Holmstrom, B. 1982. Moral hazard in teams. *Bell Journal of Economics* 13(2): 324–340.

Hoover-Dempsey, K., Battiato, A.C., Walker, J.M.T., Reed, R.P., DeJong, J.M. & Jones, K.P. 2001. Parental involvement in homework. *Educational Psychologist* 36(3): 195–209.

Ingersoll, R.M. 2001. Teacher turnover, teacher shortages and organization of schools. *American Education Research Journal* 38(33): 499–534.

Irfan, M.S. & Nawaz, K. 2012. Factors affecting students' academic performance. *Global Journal of Management and Business Research* 12(9): 17–22.

Jacobson, S. 2011. Leadership effects on student achievement and sustained school success. *International Journal of Educational Management* 25(1): 33–44.

Jagero, N. 2011. An evaluation of school environmental factors affecting performance of boarding *African Journal of Education and Technology* 1(1): 127–138.

Jeynes, W.H. 2002. Examining the effects of parental absence on the academic achievement of adolescents: The challenge of controlling for family income. *Journal of Family and Economic Issues* 23(2): 56–65.

Johnson, N.A. 1988. Perceptions of effectiveness and principals' job satisfaction in elementary schools. *Doctoral dissertation*, University of Alberta, Edmonton, Canada.

Kettle, D.P. 1997. Companionate leadership: Opening the delicate doors of the heart. *Trust for Educational Leadership* 26: 35–37.

Koontz, W. & Weihrich, H. 1998. *Essentials of Management*. New Delhi: McGraw-Hill.

Kouzez, J.M. & Posner, B.Z. 1995. *The Leadership Challenge: How to Keep Getting Extraordinary Things Done in Organizations* (2nd ed.) San Francisco: Jossey-Bass.

Kraimer, M.L. & Wayne, S.J. 2004. An examination of perceived organizational support as a multidimensional construct in the context of an expatriate assignment. *Journal of Management* 30(2): 209–237.

Kythreotis, A., Pashiardis, P. & Kyriakides, L. 2010. The influence of school leadership styles and culture on students' achievement in Cyprus primary schools. *Journal of Educational Administration* 48(2): 218–223.

Lazear, P. & Rosen, S. 1981. Rank-order tournaments as optimum labour contracts. *Journal of Political Economy* 89: 841–864.

Lee, J. 2008. Effects of leadership and leader-member exchange on innovativeness. *Journal of Managerial Psychology* 23(6): 670–687.

Liden, R.C. & Maslyn, J.M. 1998. Multidimensionality of leader-member exchange: An empirical assessment through scale development. *Journal of Management* 24(1): 43–73.

Liden, R.C., Sparrow, R.T. & Wayne, S.J. 1997. Leader-member exchange theory: The past and potential for the future. In G.R. Ferris (ed.), *Research in Personal and Human Resources Management* (15): 47–119. Greenwich, CT: JAI Press.

Loewenstein, G. and Sicherman N. 1991. Do workers prefer increasing wage profiles? *Journal of Labour Economics* 9: 67–84.

Lydiah, L.M. & Nasongo, J.W. 2009. Role of the headteacher in academic achievement in secondary schools in Vihiga District, Kenya. *Current Research Journal of Social Sciences* 1(3): 84–92.

McLaughlin, K. 1994. Rent-sharing in an equilibrium model of matching and turnover. *Journal of Labour Economics* 12(4): 499–523.

Ma, X. & Klinger, D.A. 2000. Hierarchical linear modeling of student and school effects on academic achievement. *Canadian Journal of Education* 25(1): 41–55.

Madiha, S. 2012. The impact of teachers' collegiality on their organizational commitment in high and low achieving secondary schools in Islamabad, Pakistan. *Journal of Studies in Education* 2(2): 130–156.

Maehr, M & Anderman, E. 1993. Reinventing schools for early adolescents: Emphasising task goals. *Elementary School Journal* 93(5): 593–609.

Maforah, T.P. & Schulze S. 2012. The job satisfaction of principals of previously disadvantaged schools: New light on an old issue. South African Journal of Education 32: 227–239.

Manjulika, K., Gupta, K.A. & Rajinder, K. 2008. Women in management: A Malaysian perspectives. *Women in Management* 13(1): 11–18.

Marks, H.M & Louis, K.S. 1997. Does teacher empowerment affect the classroom? Teachers' work and students experiences in restructuring schools. *American Journal of Education* 107(4): 532–575.

Meyer, J.P., Paunonen, S.V., Gellatly, I.R., Goffin, R.D. & Jackson, D.N. 1989. Organizational commitment and job performance: It's the nature of the commitment that counts. *Journal of Applied Psychology* 74: 152–156.

Midgley, C. 1993. Motivation and middle level schools. In P. Pintrich & M. Maehr (eds), *Advances in Motivation and Achievement in the Adolescent Years*. Greenwich CT: JAI Press.

Midgley, C., Anderman, E. & Hicks, L. 1995. Differences between elementary and middle school teachers and students. *Journal of Early Adolescence* 15: 389–411.

Mondy, R.W. & Robert, M.N. III. 1981. *Personnel: The Management of Human Resources*. Boston, MA: Allyn and Bacon.

Mitchell, D.E. & Collom, E. 2001. *The Determinants of Student Achievement at the Academy for Academic Excellence*. CA: School of Education, University of California.

Mudalia, A.M. 2012. The impact of head teachers' administrative factors on performance in secondary school science subjects in Eldoret Municipality, Kenya. *Journal of Emerging Trends in Educational Research and Policy Studies* 3(4): 514–522.

Murphy, J.M. 1987. Barriers to implementing the instructional leadership role. *The Canadian Administrator* 27(3): 1–9.

Musera, G., Achoka, J.K.S. & Mugasia, E. 2012. Perception of secondary school teachers on the principals' leadership styles in school management in Kakamiga Central District, Kenya: Implications for vision 2030. *International Journal of Humanities and Social Science* 2(6): 111–119.

Naseer, A.S. 2011. Successful leadership practices of head teachers for school improvement. *Journal of Educational Administration* 49(4): 414–432.

Orora, J.H.O. 1997. *Beyond the Letter of Appointment: Essays on Management (School Managers' Version)*. Nairobi: Kerabu Services Ltd.

Pamela, E. & Davis-Kean, P. 2005. The influence of parent education and family income on child achievement: The indirect role of parental expectations and the home environment. *Journal of Family Psychology* 19(2): 294–304.

Parelius, R.J. & Parelius, A.N. 1987. *Sociology of Education*. Upper Saddle River, NY: Prentice Hall.

Piore, M.J. & Doeringer, P. 1971. *Internal Labor Markets and Manpower Adjustment*. New York: D.C. Heath and Company.

Pounder, D.G., Ogawa, R.T. & Adams, E.A. 1995. Leadership as an organization-wide phenomenon: Its impact on school performance. *Educational Administration Quarterly* 31: 564–588.

Ralph, J. & Fennessey, J. 1983. Science or reform: Some questions about the effective schools model. *Phi Delta Kappa* 64(10): 689–694.

Renlhan, F. I. & Renlhan, P.J. 1984. Effective schools, effective administration, and institutional image. *The Canadian Administrator* 24(3): 1–6.

Riketta, M. & van Dick, R. 2005. Foci of attachment in organizations: A meta-analysis comparison of the strength and correlates of work-group *versus* organizational commitment and identification. *Journal of Vocational Behavior* 67: 490–510.

Robbins, S.P. 1982. *Personnel: The Management of Human Resources*. Englewood Cliffs, NJ: Prentice-Hall.

Rockoff, J., Jacob, B., Kane, T. & Staiger, D. 2011. Can you recognize an effective teacher when you recruit one? *Education Finance and Policy* 6(1): 43–74.

Rosen, S. 1982. Authority, control and the distribution of earnings. *Bell Journal of Economics* 13: 311–323.

Rowe, W.G., Morrow, J.L.J. & and Finch J.F. (unpubl.) Accounting, market and subjective measures of firm performance: Three sides of the same coin? Paper presented at the Academy of Management Annual Conference, Vancouver, 1995.

Salop, J.K. & Salop, S.C. 1976. Self-selection and turnover in the labour market. *Quarterly Journal of Economics* XC 4: 619–627.

Sammons, P., Gu, Q., Day, C. & Ko, J. 2011. Exploring the impact of school leadership on pupil outcomes. *International Journal of Educational Management* 25(1): 83–101.

Scott, W.R. 1977. *Effectiveness of Organisational Effectiveness Studies: New Perspectives on Organisational Effectiveness*. San Francisco, CA: Jossey-Bass.

Singh, S., Darwish, T.K., Costa, A.C. & Anderson, N. 2012. Measuring HRM and organisational performance: Concepts, issues and framework. *Management Decision* 50(4): 651–667.

Smith, D.M. & Holdaway, E.A. 1995. Constraint on the effectiveness of schools and their principals. *International Journal of Educational Management* 9(5): 31–39.

Somech, A. & Wenderow, M. 2006. The impact of participative and directive leadership on teachers' performance: The intervening effects of job structuring, decision domain, and leader-member exchange. *Educational Administration Quarterly* 42(5): 746–772.

Tharenou, P. 1993. A test of reciprocal causality for absenteeism. *Journal of Organizational Bahavior* 14(3): 269–287.

Wan, H.S. & Jamal, N.Y. 2012. Principal leadership styles in high-academic performance of selected secondary schools in Kelantan Darulnaim. *International Journal of Independent Research and Studies* 1(2): 57–66.

Wells, C. & Feun, L. 2007. Implementation of learning community principals: A study of six high schools. *NASSP Bulletin* 91(2): 141–160.

Wichenje, K.M., Simatwa, E.M.W., Okuom, H.A. & Kegode, E.A. 2012. Human resource management: Challenges for head teachers in public secondary schools in Kenya, a case study of Kakamega East District. *Educational Research* 3(2): 159–171.

Appendix: Students and Supervisors

Student	Brunel Supervisor	Ahlia Supervisor
Hassan A.M. Hussain	Dr Nevine El-Tawy	Dr Gagan Kukreja
Layla F. Alhalwachi	Dr Savita Kumra	Dr Samia Costandi
Litty M. Shibu	Dr Tillal Eldabi	Dr Akram Jalal-Karim
Maitham Al Oraibi	Dr Tillal Eldabi	Dr Samia Costandi
Mohammed Al-Tahous	Professor Vishanth Weerakkody	Dr Masoud Jahroumi
Najma G.R. Taqi	Dr Lynne Baldwin	Professor Wajeeh Elali
Osama F.A. Al Kurdi	Dr Ahmad Gonheim	Professor Amer Al Roubaie
Salah H. Al Hasan	Professor Francesco Moscone	Dr Gagan Kukreja
Salman A. Shehab	Dr Afshin Mansouri	Dr Jamal Zayer
Subhashini Bhaskaran	Dr Kevin Lu	Professor Mansoor AlAali
Tahani H.A. Maki	Dr Satwinder Singh	Professor Wajeeh Elali

Author index

Al Aali, M. 105
Al Hasan, S.H. 89
Al Kurdi, O.F.A. 75
Al Oraibi, M. 43
Al Roubaie, A. 75
Alhalawachi, L.F. 15
Al-Tahous, M. 55

Bhaskaran, S. 105

Costandi, S. 15, 43

Elali, W. 117
Eldabi, T. 31, 43, 117

Ghoneim, A. 75

Husain, H.A.M. 1

Lu, K. 105

Maki, T.H.A. 117

Rajab, E.K. 31

Shehab, S. 99
Shibu, L.M. 31
Singh, S. 117

Taqi, N.G.R. 67